Energy Engineering

K.V. Raghavan · Purnendu Ghosh
Editors

Energy Engineering

Proceedings of CAETS 2015 Convocation on Pathways to Sustainability

Springer

Editors
K.V. Raghavan
Indian National Academy of Engineering
Gurgaon, Haryana
India

Purnendu Ghosh
Birla Institute of Scientific Research
Jaipur, Rajasthan
India

ISBN 978-981-10-3101-4 ISBN 978-981-10-3102-1 (eBook)
DOI 10.1007/978-981-10-3102-1

Library of Congress Control Number: 2016957491

Printed on acid-free paper

This Springer imprint is published by Springer Nature
The registered company is Springer Nature Singapore Pte Ltd.
The registered company address is: 152 Beach Road, #22-06/08 Gateway East, Singapore 189721, Singapore

Foreword

I am delighted that the Post Proceedings of the International Council of Academies of Engineering and Technological Sciences (CAETS) 2015 Convocation on "Pathways to Sustainability: Energy, Mobility and Healthcare Engineering" held in New Delhi, India on 13–14 October, 2015, are being brought out. CAETS, being a nongovernmental international organization of Academies of Engineering from 26 countries, which has the broad objective of advising governments and providing an international forum for addressing all issues of concern pertaining to engineering and technology, plays a vital role in promotion of the engineering profession. Indian National Academy of Engineering (INAE), a Member Academy of CAETS, is proud to have had the opportunity of hosting the CAETS 2015 Annual Meeting and Convocation at New Delhi, India.

The CAETS event brought together eminent engineering luminaries and bright young engineers from 24 countries on a single platform who shared their country-specific knowledge on the themes of the Convocation. The CAETS 2015 Convocation focussed on the topics of Energy, Mobility and Healthcare Engineering which are of topical interest all over the world in recent times. Energy security, provision of economical modes of transport and mobility and affordable and high quality facilities for health care are necessary for the citizens of the world in order to lead comfortable and secure lives. It is indeed apt that the Convocation addressed key issues in the fields of Energy, Mobility and Healthcare Engineering that play a vital role in the lives of mankind.

It is a matter of pride that the Convocation had representation of eminent engineers from academia, R&D and industry from over 24 countries from across the globe who shared their valuable experience and expertise and enriched the deliberations with their thought-provoking presentations. These presentations were indeed brilliant and stimulating and have resulted in interesting papers which are showcased in the Post Proceedings. I thank all the authors for devoting their precious time for contributing papers for the proceedings.

I express my sincere thanks to the editors of the CAETS 2015 Convocation Post proceedings for their untiring efforts at every stage of planning of the event and in publication of the Post Proceedings. The assistance and initiative taken by Springer in bringing out the publication is deeply appreciated.

On behalf of the Indian National Academy of Engineering (INAE), it is my proud privilege to express my deepest gratitude to the CAETS organization for giving us an opportunity to host this event of international repute for the first time in India.

I sincerely hope that the event left many pleasant and memorable memories for all the delegates and am certain that the Post Proceedings of the CAETS 2015 Convocation on "Pathways to Sustainability: Energy, Mobility and Healthcare Engineering" shall be well accepted in the Engineering realm and by the Engineering community.

Dr. B.N. Suresh
President, INAE
Honorary Distinguished Professor
Former Member, Space Commission
Former Director, Vikram Sarabhai Space Centre, Trivandrum
Founder Director, Indian Institute of Space Science and Technology
ISRO Headquarters, Antariksh Bhawan
Bangalore, India

Preface

Preamble

The CAETS CONVOCATION-2015, hosted by the Indian National Academy of Engineering (INAE), was held on 13–14 October, 2015, at New Delhi, India. The CAETS Governing Council had chosen the theme on sustainable engineering pathways for energy, mobility and healthcare sectors for the conference attended by 250^+ delegates comprising of the fellowship of CAETS member academies, senior engineering experts, scientists, thought leaders, policy makers and industry executives from Belgium, Canada, China, France, Germany, Hungary, India, Japan, Netherlands, South Africa, Spain, Sweden, Switzerland, UK, USA, Uruguay and other countries. This international event was formally inaugurated by Dr. Harsh Vardhan, Honourable Minister of Science and Technology, Government of India, and he assured that the recommendations of the CAETS Convocation will be acted upon by the government. Prof. Asutosh Sharma, Secretary, Department of Science and Technology, Government of India addressed the gathering and highlighted the importance of the selected themes for developing nations.

The two unique features of CAETS-2015 were the special gesture made by the INAE (a) to provide 50 % international travel support and complimentary accommodation to encourage young engineers below 45 years of age to deliver lectures in the technical sessions along with eminent senior engineers and (b) to organize poster sessions and awards to encourage very young engineers and research scholars from CAETS member academies to present their research work. These initiatives paid rich dividends by attracting the participation of 30^+ young engineers in CAETS Convocation. Another major initiative is the formal launching of the electronic version of the CAETS report on "Transitioning to Lower Carbon Economy: Technological and Engineering Considerations for Building and Transportation Sectors" during the inaugural function of CAETS CONVOCATION-2015 by the Chief Guest. This is a joint initiative of 12 CAETS member academies from Australia, Canada, China, Germany, India, Japan, Korea, South Africa, Switzerland and UK. The experts from these academies met at Cape town (South Africa), Beijing (China), Berlin (Germany) and London (UK) during 2013–15 to identify technological and engineering priorities and emerging options for transition to lower carbon economy in building and transportation sectors.

Recommendations

Energy

Global energy consumption will grow from 524 to 820 quadrillion BTU between 2010 and 2040 leading to GHG emission increase by more than 40 % from the present level under business as usual scenario. Since the dynamics of energy growth is driven by the population and wealth enhancements, it is important to keep track of these growth dynamics in different countries. Energy resources sustainability as well as their cost efficiencies would dictate as rapid shift as possible to lower carbon nonfossil energy systems. Since under this context, recent developments in three stage thorium-based nuclear energy, large capacity solar thermal power plants, fuel cell-based independent propulsion systems for strategic and civilian applications, wind generated hydrogen as a renewable energy carrier and unconventional gas from coal seams, underground coal gasifications and biohydrogen from waste resources provide potential lower carbon energy options. A joint voluntary initiative by CAETS member academies from Argentina, Australia, Canada, China, Germany, South Africa, Switzerland and UK on unconventional gas involving basic and social scientists, engineers and economists on unconventional gas provides a new model for CAETS member academies to emulate in the future.

Developing energy scenarios under various application modes is a powerful mechanism for devising future energy-efficient systems. A typical example is the highly interconnected electricity, transport and heating sectors. The coal-based energy continues to attract engineers to develop cleaner process options and minimize GHG emissions keeping in mind the transitional priorities. The actions include upgradation of beneficiation, ultra supercritical combustion and integrated gas combined cycle concept for gasification and co-firing of coal and biomass. China's achievements in some of these areas are noteworthy. In energy materials, in which significant developments are taking place worldwide, we have to look beyond silicon for next generation solar cells by adopting conjugate organics, inorganic quantum dots and mixed semiconductor oxides/peroxides. Similar challenges are foreseen in high energy density battery and membrane materials for fuel cells. Light-based technologies can play a major part in developing future energy-efficient systems. There is a need to integrate photonic and biophotonic technologies. In this international year of light technologies, such initiatives have to be taken to achieve multi-functionality, tunability, low power for fast operations, reconfigurability and cost-effectiveness so much necessary for energy, mobility and healthcare sectors.

The transition to lower carbon regime require a careful assessment of current energy and emission loads at national level for setting energy/emission reduction targets and timelines. Selection of cost-effective technology and engineering options and their incentivization for creating new markets are the major challenges. The expert panel, which deliberated on these issues, stressed the need to develop high strength light weight materials including nano composites and nano structured steels and bio, electrochemical and catalytic options for hydrogen generation adaptable for building and transportation sectors. CAETS need to work closely with related international bodies such as World Economic Forum, International Panel on Climate Change and allied agencies.

Mobility

The future engineering challenges lie in the development of semi- and fully automated transport vehicles, intelligent driver–vehicle interface and innovative multi-axle hydraulic trailers for heavy loads. The fast changing digital technology has opened up new avenues for electrifications of rail, air and marine transport vehicles. The five digital forces, viz. cloud computing, mobile technologies, social networks, big data and robotics, will make high impact on these developments.

Exciting developments are taking place in bridge design and construction technologies with urban and rural transport restructuring, virtual mobility and carbon foot print minimization during their construction as defining factors. Novel retrofit technologies are needed for their restoration. A life cycle approach embracing bridge design, construction, maintenance, restoration and dismantling is a future priority area for civil engineers.

The recent developments in rapid urban rail transportation systems in China, India and Japan have demonstrated the new engineering skills acquired in the planning, design and execution of underground tunnels and structures. From material engineering point of view, material recycle and functionalized material applications are high priority areas in evolving sustainable roadways.

The dilemmas in mass transportation in emerging economies are many. It is important to consider seamless connectivity, smart mobility, enforcement of advanced safety measures, equitable allocations of road space for multiple type of vehicles and time variant traffic demands.

Health Care

Multitude of engineering challenges await healthcare sector in terms of new diagnostic tools, next generation medical devices and application of informatics and analytics. The recent advances in nano and point of care diagnostic, scalable medical and remote neonatal monitoring systems have enhanced the chances of early and accessible diagnosis. Structural process concept has greatly helped in developing groundbreaking technologies for these systems in advanced nations with the help of multi-disciplinary teams comprising engineers, product designers, business analysts and clinicians.

The concept of regenerative engineering which combines tissue engineering, material science, cell physics and developmental biology has enabled the technologies for next generation medical devices. Groundbreaking bionic ear and eye technologies have enabled notable progress against hearing and visual impairments. The recent advances in sensors, telecommunication and mobility engineering will play a major role in evolving next generation device technologies for both communicable and noncommunicable diseases. Sensing and data analytic skills provide new transformational material opportunities in intensive health care. Advanced computer aided tools based on big data analytics are needed for biologically meaningful insights into the enormous volume of microbiome data generated from sequencing platforms.

A panel of experts examined the issue of convergence of engineering and healthcare sciences. They are certain that the convergence is occurring impressively due to application of ICT and big analytics in healthcare systems. Several key issues including affordability of well engineered systems in rural and urban environments and engineers role in new drug discovery came up for discussion.

The major takeaways from CAETS Convocation are many. Achieving long-term sustainability under business as usual mode is virtually impossible in energy, mobility and healthcare sectors. Major emphasis has to be placed on energy consideration, expanded use of nonfossil primary energy, decarbonisation of existing energy sources and enhancing the energy efficiency of individual systems. Higher investment in R&D and demonstration is essential for technologies which are close to market maturity and those requiring scale up. Commitment from policy makers, different stakeholders, academic community and market leaders is very essential for the commercial realization. Ethical practices in Engineering are vital for achieving high level of successes in every application field particularly so in creating human artefacts that are nonexistent in nature. Gender enhancement in engineering design endeavours is essential for developing rational systems. Younger Engineers have to be given increased responsibilities in evolving innovative engineering systems. The demand for novel engineering solutions will increase exponentially while developing smart energy, mobility and healthcare systems. Research, prototyping and technology transfers in such systems have to be pathbreaking in nature. The CAETS Engineering Community is committed to create a conducive environment for open access information sharing in the above sectors between the member academies.

Gurgaon, India K.V. Raghavan
Jaipur, India Purnendu Ghosh

Contents

About the Editors

K.V. Raghavan received his B.Tech from Osmania University in 1964 and a MS and Ph.D. from the Indian Institute of Technology (IIT), Madras, India. He is a fellow of the Indian National Academy of Engineering (INAE), the Indian Institute of Chemical Engineers (IIChE) and the A. P. Akademi of Sciences (APAS) and is a distinguished fellow of the University of Grants Commission (UGC). Dr. Raghavan took up the distinguished professorship of the INAE at the Indian Institute of Chemical Technology, Hyderabad, in 2008. He became the vice president (International Cooperation) of the INAE in 2011.

He has held various levels of scientific positions in three national laboratories of the Council of Scientific and Industrial Research (CSIR) of India. He was appointed as the Director of Central Leather Research Institute, Chennai in 1994, and in 1996, he took over the Directorship of the Indian Institute of Chemical Technology, Hyderabad. On successful completion of this tenure, he was appointed as the Chairman of Recruitment and Assessment Centre of DRDO, Ministry of Defence, Government of India, in May 2004. In addition, Dr. Raghavan chairs scientific committees of public and private sector bodies on clean technologies, catalytic ethanol reforming, biorefining and allied areas. The recipient of six national awards, Dr. Raghavan has published more than 150 papers, filed 45 patents and edited five books. Chemical process development and design, reaction engineering, simulation and modelling and chemical hazard analysis are his areas of specialization. His basic research contributions cover the simulation of complex reactions in fixed bed reactors, hydrodynamics of multi-phase reaction systems, envirocatalysis for clean processing, zeolite catalysis for macromolecules, thermochemistry and kinetics of charge transfer polymerization and modelling of chemical accidents. His current research activities are in process intensification of water–gas shift reaction, catalytic CO_2 decomposition, analysis of CO_2 capture technologies and characterization of the reactivity of ionic liquids.

Purnendu Ghosh is the Executive Director, Birla Institute of Scientific Research, Jaipur. Prior to this, he was Professor at Indian Institute of Technology, Delhi. He worked at the Swiss Federal Institute of Technology (ETH), Zurich, Switzerland;

University of Melbourne, Australia; National Research Centre for Biotechnology (GBF), Braunschweig, Germany. He has taken a lead role in augmenting bioprocess engineering and biotechnology activities in the country as a member of several task forces of the Department of Biotechnology, Government of India. He is currently the Chief Editor of Publications of Indian National Academy of Engineering. He is a columnist of a leading newspaper group. Besides technical publications, he has written books of different genre.

Low Carbon Pathways for India and the World

Anil Kakodkar

Abstract Despite global consensus on the urgent need to limit global warming, firm actions to cap and reverse the level of greenhouse gases in earth's atmosphere are still eluding us. The large scale developmental needs of the developing world and the little carbon space that is available to accommodate them in the business as usual mode clearly requires immediate actions on facilitation and adoption of low carbon pathways by both the developed and the developing countries. In the Indian context, it appears that meeting the anticipated very large growth in energy needs using energy re-sources available within the country, in a sustainable way, would also require adoption of non-fossil energy pathways on a relatively urgent basis. The presentation discusses some approaches towards ensuring sustainable energy supply for meeting development aspirations of a large developing country like India through non-fossil means.

Keywords Nuclear energy · Solar energy · Non-fossil energy · Sustainable energy supply

1 Introduction

International Energy Agency's recently released report, "World Energy Outlook 2015" projects that in the new policies scenario (the central scenario), the global energy demand would grow by nearly one-third between 2013 and 2040, with all of the net growth coming from non-OECD countries and OECD demand ending 3 % lower. Despite signs that a low-carbon transition is underway, energy-related CO_2 emissions are projected to increase to 36.7 Gt in 2040, 16 % higher than in 2013. Climate pledges for COP21, if fully realized including through projected investment needs of \$13.5 trillion in low-carbon technologies and efficiency by 2030, could

A. Kakodkar (✉)
INAE Satish Dhawan Chair of Engineering Eminence, Bhabha Atomic Research Centre, Trombay, India
e-mail: kakodkaranil@gmail.com

K.V. Raghavan and P. Ghosh (eds.), *Energy Engineering*,
DOI 10.1007/978-981-10-3102-1_1

possibly mean a temperature rise of 2.7 °C [1]. The efforts thus are not yet enough to move the world onto a pathway consistent with the 2 °C climate goal.

Current total Indian energy use stands at around 0.8 Btoe per year. India's energy demand during this period would see a large rise driven by new infrastructure, an expanding middle class and 600 million new electricity consumers. By 2040, India would close in on United States in terms of energy demand even though demand per capita would remain 40 % below the world average [1]. India is set to contribute more than any other country to the projected rise in global energy demand, around one-quarter of the total. To reach a human development index (HDI) comparable with that in advanced countries (~0.92), India's energy consumption needs to rise to around 4 Btoe per year. Although the total primary energy resources (coal, oil and gas) in India are estimated at around 360 Btoe, actual minable resource (particularly coal) is likely to be much less. The desirable level of energy use to sustain a decent human development index (~0.9+) is unlikely to be sustained (see Table 1) on the basis of available fossil energy resources in the

Table 1 Indian energy resources

Sr. No.	Parameter	Total	Per capita	Remarks
1	Current (2015) energy use per year	0.820 Btoe	640 Kgoe	Most of oil and some gas and coal imported. Import of energy likely to rise steeply (>5–7 times in 15–20 years)
2	Desirable energy use per year	4.125 Btoe	2500 Kgoe	To reach HDI of 0.9+
3	Total primary energy reserves (coal, oil, gas)	361.8 Btoe		Actual mineable is much less. Need technology to make better use of available resource
4	Total renewable potential (incl. hydro, excl. solar) 0.2 Btoe/year	~300,000 MW (peak)		~5 % of requirement
5	Solar resource	~45,000 km^2 which corresponds to a fourth of barren and uncultivable land in India would be sufficient to meet entire electricity requirements (~20 % of total energy requirement)		Needs emphasis on solar thermal (for both electricity and hydrogen)
6	Nuclear resource	Uranium ~87 Btoe Thorium >600 Btoe		On the basis of nuclear recycle

country for a long enough time. Already India imports nearly a third of its energy requirement. Projections are that the energy imports would grow by 5–7 times in next 15–20 years mostly in the form of fossil energy. This clearly would have serious implications in terms of balance of payments. Not only the import bill in absolute terms would be large but the implications of potential volatility would be even larger. Thus quite apart from concerns related to climate change and energy resource sustainability, the implications related to large scale import of fossil energy would also dictate a rather rapid shift to non-fossil energy.

2 Non-fossil Energy

The Table 1 above which has been put together on the basis of available data from different sources also indicates that renewable energy excluding solar but including hydro would constitute about 5 % of the total requirement of 4 Btoe per year. Thus while all these energy sources are important in the short run, we need to aggressively develop solar energy and nuclear energy, both of which are available on the Indian land mass in abundant measure, to ensure sustainable high quality of life of our people.

India's 'intended nationally determined contribution' (INDC) announced before Paris Conference (CoP 21) clearly states the intention to reduce the emissions intensity of its GDP by 33–35 % from 2005 level and to achieve about 40 % cumulative electric power installed capacity from non-fossil fuel based energy resources, by the year 2030, with the help of transfer of technology and low cost international finance, including from Green Climate Fund. An assessment of our current non-fossil electricity generation indicates a generation capacity of 81,757 MWe as on August 2015. By the year 2030, the electricity generation capacity is likely to be around 746 GWe [2]. Thus around 215 GWe non-fossil energy based generation capacity would need to be added in next fifteen years. Government of India has already set a target of generating 175 GWe of renewable energy by the year 2022. Taking this and the expected contributions from nuclear and hydro sector into account, there should be no difficulty in realising the non-fossil energy based generation target as spelt out in our INDC. Looking at the larger problem discussed earlier, we could perhaps do much more.

A question may be raised on the need to pursue both solar and nuclear energy. In this context, it is important to recognise the need to have a basket of energy supply with built in diversity. As it is, in the long run, we have only two options as of now. Besides the two sources have different but complimentary features. One available 24 × 7, the other intermittent. One suitable for decentralised generation, the other for central generation. To en-sure stable operation of the grid in a cost effective way, there should be a minimum base load generation which can more easily be from nuclear plants while distributed solar generation closer to load can help regulation. Concurrent use of both solar and nuclear is therefore inevitable.

There are however challenges that we need to recognise and overcome. For large scale deployment both solar and nuclear energy, establishment of competitive domestic supply chain is a must. In case of nuclear energy, the technology for manufacture of required materials and equipment has been developed for reactor systems like PHWRs and FBRs developed locally. We need to be doing similar thing for reactors based on imported technology. Thus while availability of technology may not be a big issue, the program deployment rate is rather slow. On the other hand, in case of solar, the deployment rate has really picked up but there are difficulties in sustaining manufacturing supply chain with basic raw material manufacture in the country practically non-existent. Since large investments are involved in such programmes, it is important that good part of that money is spent in the country to support local industry and related job creation.

3 Solar Energy

While deploying solar energy, we must invariably attend to specific Indian requirements such as high temperature, high dust level, water scarcity, possibility of abrasion in case of installations located in desserts, uneven terrain in case of installation located in hilly areas and mountains, (both these location choices are otherwise attractive in view of high insolation and relative freedom from land use conflicts) etc. There is also merit in revisiting DC end use appliances running on photovoltaic sources because of their significant potential in terms of better economics as well as better energy efficiency. Cost effective energy storage continues to be the biggest challenge in large scale penetration of solar power.

In case of solar thermal technology, it's potential for competitive 24×7 electricity generation through large capacity plants has not yet been fully recognised. Further, solar thermal technology is inevitable for realising the role of solar energy as primary energy that can be used for both production of electricity as well as for pyro-processes including for production of hydrocarbon substitutes or even hydrogen. An added advantage with solar thermal technology is the possibility of near full indigenisation relatively easily. There is also the possibility to realise viability even with small solar thermal installations with the developments like Brayton cycle and related equipment.

In order to demonstrate solar thermal power generation technologies on MWe scale from where credible extrapolations for viability of large capacity plants can be made, two projects are currently being worked upon. BARC and ONGC are developing a 2 MWe solar thermal beam down facility. Such a facility, since its receiver furnace is located on ground, can also be used to explore the use of solar energy for other pyro-chemical or pyro-metallurgical processes. The other project is being developed by IIT Bombay and NTPC for setting up a 3.5 MWe plant capable of round the clock operation.

Development of solar technology products consistent with specific Indian needs including technologies for their manufacture needs much greater attention. An important strategy could be to promote such technology product development in a demand driven mode by selecting and supporting a few of proposals made jointly by an industry and a laboratory against the requests issued for the purpose. We also need to develop policies that encourage market entry of duly qualified products so developed.

4 Nuclear Energy

The three stage nuclear power programme of our country is well on its way in technological terms. The international embargos that the programme was facing are largely gone without our having to compromise on our strategic autonomy. While we continue to pursue further advancing the technology, the need of the hour is to accelerate programme deployment overcoming the barriers faced. While large scale thorium utilisation remains the key long term objective, there are also possibilities of a more rapid global deployment of nuclear energy leveraging some of the attributes of thorium [3]. Use of thorium matrix fuels in most of existing reactors could well address the concerns related to nuclear proliferation as well as safety and long term waste management without adversely impacting on energy use and economy. India with her advancement in the area of thorium is well placed to take a lead in this context.

Accelerator driven sub-critical reactor systems have the potential to efficiently transmute long lived radio-active species and also permit some growth with thorium fuelled systems. This along with partitioning technologies could also virtually eliminate the long term waste problem. In addition to the technologies related to the three stages of our nuclear power programme, we also need to develop technologies for use of nuclear energy as primary energy source providing high temperature to permit pyro processes as mentioned earlier. We need high temperature reactors for this purpose. Development on these fronts is well underway.

5 Role of Solar and Nuclear Energy as Primary Energy

Apart from developing these two primary energy resources for production of electricity, a major thrust would be needed to produce non-fossil hydrocarbons/hydrogen from these primary resources, so that all segments of energy demand can be addressed. Thus apart from building high temperature capability in solar and nuclear energy technologies, one would require several other technologies such as for production of hydrogen in a more economical way, use of bio-mass and hydrogen for production of hydrocarbon substitutes for use in transportation and other sectors, possibility of recycle of CO_2 through biomass and other modes [4] of CO_2 sequestration, appliances/equipment that can use alternate fluid fuel forms such as hydrogen and a host of such other technologies. Following

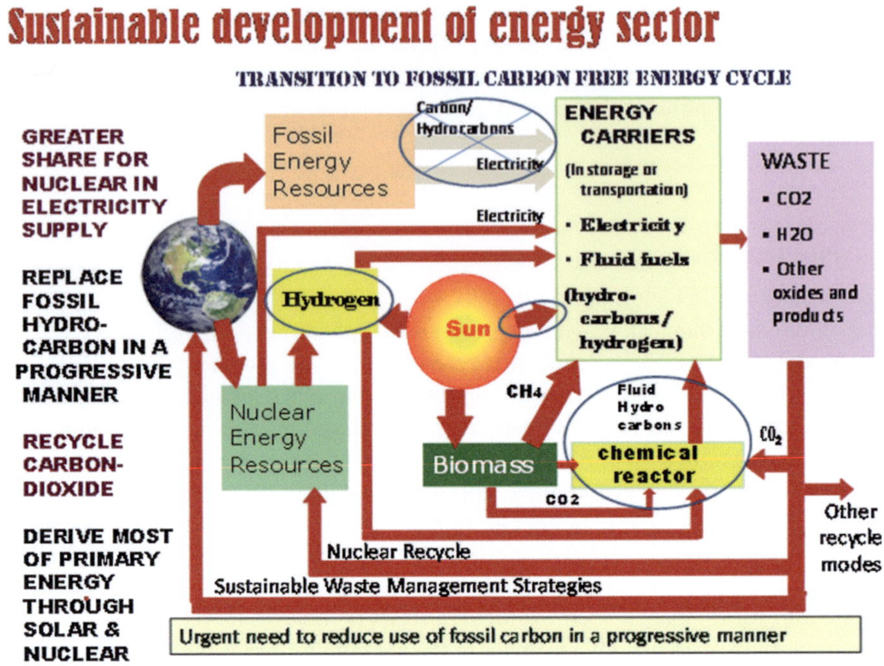

Fig. 1 Transition to fossil carbon free energy cycle

such a strategy, one can indeed make a transition to nonfossil carbon energy cycle, a matter of crucial importance to limiting the global warming. Such a programme needs to be pursued with a time bound mission mode approach (Fig. 1).

6 Concluding Remarks

India's energy challenges are very large with ramifications not just for India but for the whole world. Fortunately the challenge of climate change, sustainability and finance all point towards a common solution. However intense R&D in a mission mode is necessary to develop appropriate solutions and put them in place.

References

1. World Energy Outlook (Executive Summary, Fact Sheet, Presentation to Press), International Energy Agency, 2015
2. India Energy Outlook—World Energy Outlook special report, International Energy Agency 2015

3. Anil Kakodkar, in *Towards Sustainable Secure and Safe Energy Future Leveraging*. Opportunities with Thorium" Thorium Energy Conference 2013 (ThEC 13) CERN Geneva 28–31 Oct 2013
4. C. Graves, S.D. Ebbesen, M. Mogensen, K.S. Lackner, Sustainable hydrocarbon fuels by recycling CO_2 and H_2O with renewable or nuclear energy. Renew. Sustain. Energ. Rev. **15**, 1 (2011)

Fuel Cell Technologies for Defence Applications

J. Narayana Das

Abstract By virtue of their distinct features like autonomy, low signatures, no emissions, high specific energy etc., fuel cells could find a number of applications in the defence sector. Depending on the specific context, the requirement could be for a wearable, portable or distributed power supply. Powering unmanned aerial vehicles, ground vehicles and autonomous underwater vehicles form a separate regime. High levels of efficiency, reliability, reproducibility, robustness to meet the MIL standard environmental tests etc., are the prerequisites for military hardware. The salient features of fuel cells are touched upon in this context and design approach for a fuel cell based AIP system for submarines is discussed in brief.

Keywords Fuel cells · Soldier power · Portable power · Auxiliary power · Unmanned vehicle propulsion · AIP for submarines

1 Introduction

Defence forces look forward to self sufficiency in every situation and location. Power and energy supply must be robust, reliable and versatile. Batteries of several types and specifications have been specially engineered and are in wide use by the Army for their forward area detachments and by the Air force and Navy for autonomous vehicles and remote operation fields. The concept of fuel cell demonstrated by Dr. William Grove in 1839, has undergone numerous innovative up gradations and has got adapted and diversified into several types. Low acoustic signature, low thermal signature, practically no chemical emission, improved specific energy, high energy density, reduced recharging cycle times etc., are important features of fuel cells weighing against the best of battery choices, as far as the military segment is concerned. These features are of significance to the civil sector as well. Still the industry has not been able to penetrate the market, to the

J. Narayana Das (✉)
NPOL, Trikkakara, Kochi 682021, India
e-mail: jpndas@gmail.com

K.V. Raghavan and P. Ghosh (eds.), *Energy Engineering*,
DOI 10.1007/978-981-10-3102-1_2

extent it should have; despite the global green energy campaign. According to a 2015 review, cumulative installed capacity of fuel cells since 1995 is just about 1 GW [1] and the shipment of units forecast for 2015 is around 160,000, all categories included.

Batteries and Fuel cells are both power on demand devices based on electrochemical energy conversion. In batteries the stored chemical energy is released as electrical energy as a result of reactions between the electrodes and the electrolyte. Once the reactants are consumed, the battery stops delivering power and needs to be recharged using electrical energy from external sources. In fuel cells, though electrical energy is generated through electrode reactions, the reactants per say are not stored in the cells and can continue to give rated power output as long as supply of the fuel and oxidant could be maintained.

Polymer electrolyte fuel cell (PEFC) uses a solid polymer electrolyte membrane for exchange of the H^+ ion facilitating the anode and cathode reactions of Hydrogen and Oxygen, using Platinum and alloy catalysts. Relatively lower service life, stringency of material specifications and the need for extremely high purity of Hydrogen, etc., are the limiting factors. Solid oxide fuel cells (SOFC) are made up of ceramic and cermet electrodes and electrolyte systems such as Yttria Stabilized Zirconia (YSZ). They are robust in nature. No gel or liquid or polymer membrane is involved. However, the engineering challenges are several, since the operating temperatures are as high as 800–1000 °C. Handling Hydrogen gas at such high temperatures is a safety critical issue. In stationary systems SOFC is finding large scale application, primarily due to the flexibility in fuel choice. Ni-cermet anode used in SOFCs, has very poor sulfur tolerance below 800 °C. To be successful in automobiles, PEFC system must operate at 110–120 °C, which causes associated performance and degradation issues [2].

In Phosphoric acid fuel Cell (PAFC), phosphoric acid spread over a porous supporting substrate forms the basic electrolyte layer. Platinum and platinum alloys on Carbon form the catalysts. Handling the corrosive acid, maintaining its concentration and choice of acid resistant materials are the critical engineering challenges. Though the operating temperature is higher in comparison with PEFC, better tolerance to impurities in the reactants is a specific advantage. Overall power to weight ratio is lower than that of PEFC. But PAFC has much longer service life. Alkaline Fuel Cells (AFC) have the fastest kinetics. The electrode support is typically Ni mesh or foam. The separator media is alkali (typically KOH) soaked asbestos membrane. Such systems can use metallic bipolar plates, and thereby reduce cost. However, their vulnerability to poisoning by CO_2, corrosion of the electrodes, dilution of alkali in the cell etc., are issues that restrict the use of AFC. Molten Carbonate Fuel Cells (MCFC) use fused alkali carbonates as primary electrolyte and bipolar plates are of metal alloys. Since operating temperature is high, faster kinetics is possible. Major problem encountered is the corrosion of electrodes.

In all of the above cases hydrogen fuel is supplied externally, either as stored gas/liquid, or through reforming hydrocarbons. Alternately decomposition of peroxides is also practiced as an option. In Direct Fuel Cells (DFC) hydrogen containing materials like methanol or sodium borohydride is directly used as the fuel, rather than using an external fuel processor or reformer, to generate hydrogen online. Such fuel cells are more compact though they suffer from much slower kinetics.

2 Potential Defence Applications

Simplicity, durability, ruggedness and high level of autonomy, are essential features of any military hardware. Systems should have fault diagnostics and self protection features. Detailed maintenance plans, mean time between failures (MTBF), mean time to repair (MTTR), etc. are important statistical parameters of interest to the military customer.

Air force bases in forward areas and remote locations need assured electric power for battery charging, auxiliary power for surveillance and regular power for communication equipment. Long endurance unmanned aerial vehicles also need agile power sources. Navy's strategic need of electrical power is for running the unmanned underwater vehicles and air independent propulsion systems for non nuclear submarines. The land forces cannot be confined to pre chartered fields and terrains. Power supply would have either been destroyed or had never existed in the new posts they occupy. Based on the typical operating environment and user perspectives, army's power requirements can be classified into soldier power, auxiliary power units (APU), autonomous systems, distributed power plants etc. [3]. US department of defence has carried out a comprehensive study and has identified the distinct areas, as soldier wearable and portable power, auxiliary power units for ground vehicles, ships, and aircrafts, non-tactical light-duty vehicles, propulsion power for ships, submarines, autonomous underwater vehicles (AUVs) and unmanned aerial vehicles (UAVs) [4].

2.1 Soldier Power

Portable high density power source is vital for the modern war fighter to meet his C4I system needs. Apart from the communication equipment, power is required for helmet mounted displays, mobile computer, data modems etc. Primary challenge is to keep the system weight low. High temperature PEFC with methanol reformer is a choice for 25–55 W systems. Portable model JENNY 600 S from M/s. SFC Energy can directly power electrical devices or recharge secondary batteries. The fuel cartridge contains methanol and has a capacity of 400 Wh each. Ultracell has been able to pass several models of their reformed methanol fuel cells (RMFC) through the rigorous test procedures of the US army. PEFC based prototypes developed by M/s Ballard power systems is using sodium borohydride as the primary fuel. Naval Materials Research Laboratory, (NMRL), India has developed a 100 W system based on PEFC technology, integrated with online hydrogen generator, for man portable field power applications. The system can provide 100 W power for 10 h for every 1 l of liquid fuel.

Operation at low ambient temperatures of the order of −20 °C, as well as performance at low ambient oxygen levels, typical of the high altitudes has to be specially factored in the design of fuel cells for such applications. For use in desert regions, the system should take care of high temperature autocatalysis of hydrogen donor materials employed. Similarly, if liquid fuels like methanol is used, due consideration should be given to the fuel's flash point. Air breathing systems should employ dust filters engineered to take care of the desert storms. Practically no repair is possible in the field. Systems should have long enough MTBF. Power conditioning to meet the input requirement of specific devices and qualification of system to MIL standards of EMI/EMC are other engineering challenges.

Lower end of soldier power, say around 20 W, can be met by direct ethylene glycol—anion exchange membrane based fuel cell. Ethylene glycol being an anti-freeze material is suitable for low temperature locations like northern sectors of India. Direct Methanol Fuel cells are also showing high potential for such low power applications. A wearable fuel cell together with disposable fuel cartridge can provide higher energy density than the best of the lithium primary cells.

2.2 *Auxiliary Power Units (APU)*

Field deployed vehicles and battle tanks of army need on-board power for electrical and electronic devices in use. Auxiliary power required during 'silent watch' should leave absolutely low signatures. Such systems should be capable of autonomous operation without operator intervention. The system should be engineered to give high levels of reliability under extreme environments of temperature, dust, humidity, shock and vibrations. Processes should not leave observable emissions of chemicals, smoke, light or sound. Ideally their thermal signatures should also be very low. Power conditioners and associated electrical circuitry should conform to MIL standard EMI/EMC specifications. This is a specific application where the conventional diesel power generators can be replaced with fuel cell generators for significant strategic advantages. Weight and volume considerations are important; but not as critical as in the case of man portable systems. However, the systems need to be all weather resistant, robust and highly reliable. Operator intervention and maintenance requirements should be minimal.

NMRL has developed a PAFC based 10 kW generator car that uses an integrated methanol reformer for in situ hydrogen generation. This power source can be used with advantage for ad hoc repair facilities for field equipment. They are also handy for enhancing relief operations in distress management and for providing emergency medical assistance camps in the remote locations.

2.3 Distributed Power Generation

At forward area base camps of the armed forces grid power may not be available. Captive generation is the only alternative. Major portion of base power needs, as well as heating and cooling needs, can be met by fuel cell systems. Combined heat and power (CHP) systems with fuel cells can be very effective at remote locations. Ground handling vehicles can be directly operated by fuel cells, or alternatively fuel cells can charge the battery operated vehicles and equipment. Higher overall efficiency of the system will justify the high initial cost since transportation of fuel to such locations through difficult terrains is very cumbersome and expensive. Maintenance free operation for long periods is a primary requirement. PAFC is a good choice from the life expectancy considerations. But SOFC is more versatile when choice of fuel is considered. For this segment, NMRL has developed a 30 kW truck mounted modular PAFC power plant with integrated methanol reformer.

2.4 Autonomous Systems

Advent of robotics and unmanned vehicles have revolutionised strategies and tactics in the battle field. Unmanned aerial vehicles of diverse capabilities, unmanned ground vehicles used for mine clearing, unmanned NBC reconnaissance vehicles, autonomous underwater vehicles in diverse roles etc., are the new generation technologies aimed at reducing human causalities in situations of conflict. These special platforms require widely varying power sources, typically 10 W for micro aerial vehicles at the low end and up to 3000 W for unmanned ground vehicles.

Unmanned underwater vehicles (UUVs) require power sources with long endurance having high specific energy and power density, beyond what can be met using conventional battery power. In order to realize the full potential, current research focus is on advanced batteries and mini fuel cell systems. Fuel cells can have energy density several folds higher, compared to silver-zinc or lead acid batteries. As the energy storage volume of a UUV increases, fuel cells become more appropriate for enhancement in run duration, compared to even Li-Ion batteries. Unlike batteries which take a long time to recharge, the fuel cell's tanks can be refuelled quickly, resulting in rapid turnaround times between missions. A wide spectrum of technologies starting from mini-tubular SOFC to light weight PEFCs, conformal DBFCs etc., have been specially developed with success.

Use of hydrogen as compressed gas and oxygen as liquid oxygen(LOX) has been the choice for fuel cell system on 'URASHIMA' UUV [5]. Fuel cell is housed in a pressure vessel. Oxygen gas is supplied from a high pressure oxygen gas tank and hydrogen is supplied from the metal hydride contained in a pressure vessel.

Chemical storage and onboard release of oxygen on demand from decomposition of peroxides and perchlorates offers high volumetric efficiency, though weight penalty will be higher [6]. Generation of hydrogen from hydrolysis of borohydrides, is also an available choice. For UUVs the volume constraint is important and it is necessary to minimise the balance of plant. Ease of recharging and refilling, is also one of the important criteria.

Depending on specific missions, UUVs have to dive to very deep ocean (say a few thousand meters) to retrieve data from sea bed sensor systems, or travel at medium depths (say a few hundred meters) for tracking and trailing adversary submarines. In either case the energy density as well as specific energy of the power plant has to be very high. The system should be heat integrated to the full extent, so that net heat thrown out should not increase the ambient temperature inside the electronics compartments. Being deep water vessels hull penetrations are generally avoided. All reaction products are to be contained inside the body. General arrangement of fuel cell and balance of plant should be so designed as to ensure that the progressive shift in centre of gravity and centre of buoyancy, during the run of the UUV are within the permitted hydrodynamic design limits of the platform. The system needs to be engineered to be fully operational even during the extremes of manoeuvres of the vessel. There are UUVs that are designed to run in different modes totalling to several says in a single mission. The command and control during the mission will be by the onboard computer. Power plant reliability has to be absolute, for such applications, where the plant needs to respond to the dynamic load characteristics and function without attention, for several days at a stretch.

Alejandro Mendez et al. [7] has reviewed some of the field demonstrations of autonomous vehicles powered by Fuel cells. Several models of fuel cells developed for AUVs, comparison of fuel options, oxygen storage possibilities and the associated issues are discussed in detail.

2.5 Fuel Cells on Battle Ships

The United States as well as United Kingdom have programmes running, to study the advantages of using fuel cells for powering surface ships. Specific advantages are better efficiency compared to gas turbines and diesel engines, reduced smoke, reduced sound and thermal signatures, lower vibration levels, design flexibility due to modularity etc. Power rating required will be of the order of a few megawatts. Types of fuel cells having potential utility are SOFC, MCFC, PAFC as well as PEFC. On board fuel processing will be inevitable. Use of logistic fuel is very much desired, if dual modes are proposed. Hybridization of direct fuel cells with turbine cycles using the fuel cell by-product heat, as proposed by M/s Fuel Cell energy® [8] can give very high overall efficiencies.

2.6 *Air Independent Propulsion System (AIP)*

Air Independent Propulsion system, popularly known as AIP system, for conventional diesel electric submarines, is a mission critical application for fuel cells. Diesel electric submarines use storage batteries as power source during their subsurface sailing missions. The batteries are recharged using electricity produced by running the diesel generators, for which the boat needs to snorkel. AIP systems cater for charging of the batteries while the submarine is still in the dived condition. The added feature has a force multiplier effect on the role and performance of the submarine, since the vessel can undertake very long underwater sailing missions. Typically the endurance can be up to 2 weeks, whereas a non-AIP submarine has to surface once in almost every 2 days. Of the different types of AIPs, fuel cell based AIPs are considered superior, thanks to its silent operation and low heat generation. Reliability, availability and maintainability of the system are most important for submarine applications. Modular architecture has advantages over a composite system [9]. Even if one of the modules fails, an intelligent control system can regroup and realign the healthy units to give at least a reduced output, thereby enhancing survivability.

AIP module is commonly engineered as an auxiliary standalone unit. The module has to be comprehensive and complete including provision for the fuel processing/storage, oxygen storage and dosing, control and instrumentation systems, microclimate management, safety features, as well as platform interfaces. The fuel cell per say and the balance of plant, together should geometrically conform to the form, fit and design standards of the submarine. The system should satisfy all the platform doctrines.

Structurally the submarine is designed as an externally loaded pressure hull and will be circular in cross section. The add-on AIP section should have only marginal impact on the speed and manoeuvrability of the vessel. As a rule about 10 % extra length is permitted. A typical AIP section can therefore be a cylindrical plug, about 6.5 m in diameter and about 7 m in length. The plug has to be neutrally buoyant and can therefore it can have a typical weight of the order of 200–250 tons.

2.6.1 Weight and Volume Constraints

The volume budget should consider space for passing pipelines cables and trunking through the AIP module to connect between the fore and aft sections as well as passage for crew movement. Apart from refuelling and local maintenance requirements, provision has to be made for shipping in and shipping out of equipment for shop floor repairs and refit. Submarine safety codes stipulate the porosity, i.e., the total volume of equipment in the module, expressed as a fraction of the total volume of the plug. This is a cardinal requirement for the crew safety. With the total reactants loaded, the module should be neutrally buoyant. In case the design calls for any of the by-products of onboard chemical processes to be discharged into the sea, compensating tanks for maintaining the buoyancy criteria has to be provided. The lay out design shall ensure that only minimal shifts in centre of gravity will be experienced, as the reactants get consumed.

2.6.2 Constraints in Equipment Sizing and Lay Out

Loss in energy at any stage will have cascading effects. Primarily, for the same mission endurance more amount of fuel has to be carried, increasing the all up weight. Secondly the lost energy will manifest as heat generated within the compartments leading to increased air conditioning load, which in turn adds to the parasitic load on the system. Efficiency of power conditioners should be at least 95 %. The load dependent voltage characteristics, typical of fuel cells, throw immense challenges to the power system hardware designer.

In general practice the fuel cells are stacked using stress bolts. The level of pre-stress is chosen by the designer with due considerations for the gasket seating force required for the individual cells, the contact pressure required for current collectors of the cells, thermal expansion and stress relaxation characteristics of the prestressed assembly, etc. The stacks have to be configured and engineered to form towers of optimal capacity. The preferred shape of cocooned towers is cylindrical so that the unit can be shipped out through the circular hatch openings. Being equipment onboard the submarine, all equipment designs will have to qualify the shock and vibration standards through use of appropriate shock mounts and anti vibration devices.

2.6.3 Platform Safety Concerns

Any leakage of Hydrogen inside a submarine compartment is strictly prohibited. Measures such as jacketed pipe design and special welding and joining procedures have to be adopted. Reliable leak detection and abatement systems have to be provided for. The designs should be qualified through physical tests of conformity under environmental and cyclic thermal aging processes. Hydrogen requirement for a sortie is to the tune of a few tons. In certain designs of AIP hydrogen is stored in metal hydrides and is desorbed on a temperature swing. To avert the potential hazard from accidental temperature rise and consequent high rate release of hydrogen, the hydride cylinders are located outside the pressure hull. The design calls for hull penetrations to take the gas inside. Optionally hydrogen is produced on board through chemical routes-either by hydride decomposition reactions or by partial oxidation of alcohols or through reforming of diesel. Prima facie this arrangement is safer since there is no bulk storage of hydrogen at any time. However, designing a complex chemical plant for the dynamic conditions of a submarine and engineering the plant for hydrogen safety regimes under continuous operational cycles is a challenge in itself. Behaviour of plant hardware and equipment has to be studied using ship motion simulators. Geometrically the equipment designs have to be tailored to suit the form and fit of cylindrical profile. Process intensification and reliability engineering has to be thorough. Gravity flow conditions under different list, trim and roll conditions of the submarine in motion have to be simulated while qualifying the plant architecture.

Oxygen requirement for a full sortie will be to the tune of 30–40 tons. The most efficient way to store oxygen is as liquid oxygen (LOX). LOX tanks have to be located inside the pressure hull for single hull submarines. Upper limits on thermal leakages are very critical since excessive boiling of LOX can lead to forced venting routines. Designer should ensure that the LOX system does not add to criticalities during the permitted motion cycles of the ship under various environments of shock, vibration and acceleration/deceleration.

3 Concluding Remarks

Fuel cells have come a long way in technology maturity. Still large scale exploitation in both, domestic and industrial segments has not taken place, at the expected pace. Other forms of energy conversion are still remaining competitive. Extensive R&D focused on cost reduction and life cycle cost management is progressing. Defence sector stands to gain significantly from the unique features of fuel cells. The challenges in design and engineering to meet the stringent military standards in reliability, environmental qualification, life cycle management, etc., are to be addressed through a comprehensive and holistic approach. Application development programmes should look at the totality of system and not the fuel cell alone in isolation. In this paper, certain key issues and important engineering challenges to be addressed in development of a fuel cell based AIP system for submarines, has been dealt in some detail in order to illustrate the system engineering complexity.

References

1. The Fuel Cell and Hydrogen Annual Review, 4th Energy Wave (2015), p. 4
2. S.B. Adler, Fuel cells: current status and future challenges. The BRIDGE **35**(4), National Academy of Engineering 29–32 (2005)
3. A.S. Patil, T.G. Dubois, N. Sifer, E. Bostic, K. Gardner, M. Quah, C. Bolton, Portable fuel cell systems for America's army: technology transition to the field. J. Power Sources **136**, 220–225 (2004)
4. T.J. Gross, A.J. Poche, Jr., K.C. Ennis, *Beyond Demonstration: The Role of Fuel Cells in DoD's Energy Strategy*, Chap. 2, p. 5
5. T. Maeda, K. Yokoyama, N. Hisatome, S. Ishiguro, K. Hirokawa, T. Tani, Fuel cell AUV "URASHIMA" mitsubishi heavy industries, ltd. Techn. Rev. **43**(1) (2006)
6. K.E. Swider-Lyons, R.T. Carlin, R.L. Rosenfeld, R.J. Nowak, Technical issues and opportunities for fuel cell development for autonomous underwater vehicles. AUV—Symposium on autonomous underwater vehicle technology http://www.Ieeexplore.Ieee.Org ar number=1177204

7. A. Mendez, T.J. Leo, M.A. Herreros, Fuel cell power systems for autonomous underwater vehicles: state of the art, first international e-conference on energies, 14–31 March 2014. http://sciforum.net/conference/ece-1
8. Fuel cell energy®, fuel cell power plant experience, naval applications, US department of energy/office of naval research, shipboard fuel cell workshop Washington, DC March 29, 2011
9. P.L. Mart, J. Margeridis, Fuel cell air independent propulsion of submarines, DSTO-GD-0042, Department of Defence, Defence Science And Technology Organisation Australia

Wind Energy for Buildings and Transport Sectors

S. Arunachalam

Abstract Current scenario on useful exploitation of wind and wind generated hydrogen as renewable sources of energy in buildings and transport sectors respectively is presented. Future challenges are also indicated.

Keywords Wind · Renewable energy · Buildings · Transportation sectors

1 Introduction

Wind is a useful form of renewable energy given by Mother Nature to humankind. We welcome it when it is a breeze, while we hardly prefer it when it acts as an extreme wind event such as cyclones/hurricanes, causing colossal loss of life and property. Wind being abundant and of renewable type, harnessing wind energy has always been globally considered to yield a source of clean, green, eco-friendly and sustainable energy. Wind engineering as defined by Prof. J.E. Cermak, is recognised "as the rational treatment of interactions between wind in the atmospheric boundary layer and man and his works on the surface of earth" [1]. A study on applications of wind for energy purposes is a subset of wind engineering. Currently fossil fuels play a significant role in meeting the energy demand of most of the countries in the world including India. The energy distributions from different sources of energy with current and future demands in our country are shown in Fig. 1 [2]. The thermal power using coal continues to be the dominating source of energy contributing to more than about 57 %. Within a span of about 20 years, between 1990 and 2009, the energy demand in India has nearly increased by 100 % from 320 million tons of oil equivalent (Mtoe) to about 669 Mtoe. The present

S. Arunachalam (✉)
Jaypee University of Engineering and Technology, Guna, MP, India
e-mail: sarunacha@yahoo.co.in

K.V. Raghavan and P. Ghosh (eds.), *Energy Engineering*,
DOI 10.1007/978-981-10-3102-1_3

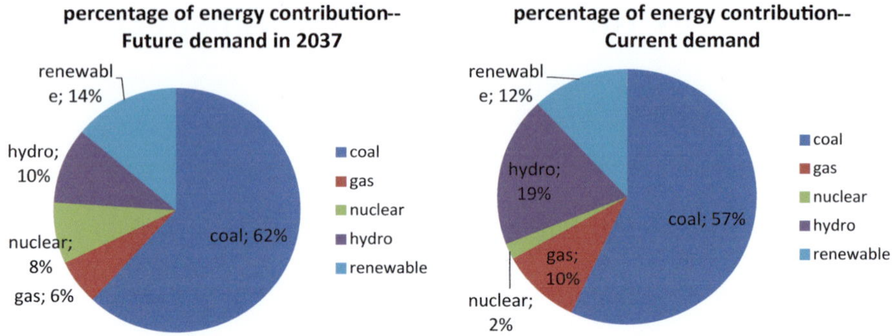

Fig. 1 Energy contribution by various sources of energy-current and future demand [2]

higher cost of import of oil and gas, and fast depletion of available fossil fuels impose severe concerns for the engineering community for developing innovative energy resource materials, focussing on related technological challenges and to provide feasible, economical and technically viable solutions which are required to satisfy the energy requirements of the country. Further, use of fossil fuels such as coal, wood, oil, natural gas, and biomass release harmful green house gases into the atmosphere polluting the environment. The Government of India emphasises through "make-in India" programme, the need to provide a clean, green, and sustainable living and environment to the society. In India, the energy consumption for building sector is about 29 %. Infrastructure with high-rise buildings is being focussed as a key area for the near future development by the Govt. of India, the constructed building space area is expected to increase to about 1900 million m^2 from the current value of 660 million m^2 [2]. Towards this the useful exploitation of wind as a promising source renewable energy, both in building and transport sectors is discussed in this paper.

The author as a researcher has been working in the area of wind engineering in investigating wind effects on various civil engineering structures such as high-rise buildings, power plant structures etc., through boundary layer wind tunnel testing, and also in conducting full-scale field measurements on wind characteristics, and wind induced response of towers, as shown in Fig. 2 [3]. In this regard, the author admits that this paper presents current scenario for the use of wind as a source of renewable energy for building sector and transport sector, including future challenges, and it is based on technical information referred from various literature and other sources including internet.

Fig. 2 Application of boundary layer wind tunnel testing on civil engineering industrial structures [3]

2 Wind Energy for Building Sector

Innovative ideas of integrating a wind turbine in conventional building design to produce electricity using wind as a renewable energy source has been demonstrated in the design and construction of Bahrain World Trade Centre (BWTC), in Manama, Bahrain [4, 5]. This project which has been planned to reduce carbon footprints and also the dependence of fossil fuels, and to increase sustainability concepts in building design, chose a new concept for using wind energy for buildings. While many of these details are available freely in open literature, for completeness of discussions, some salient details are presented here. The BWTC is a 240 m, 50 floor high twin tower complex used as a commercial building with an elliptical and sail-shape as shown in Fig. 3. Using aerodynamic principles by suitably changing the geometry of structure along the height with a taper, to reduce drag at the top, and by funneling the oncoming wind to flow between the twin towers and by parallel installing of three large wind turbines (with diameter of 29 m each with a capacity of 225 kW) at different heights, wind energy is used effectively to produce electricity. An estimated wind induced electricity generation of 1100–1300 MW per year was claimed to be made possible. This would correspond to about 11–15 % of total power consumption of the towers. For approach winds within the sector of 270°–360° with respect to center line of the twin tower, the operation of wind turbine is recommended based on wind tunnel studies. For winds outside this range, the blades would automatically be set for standstill condition. In order to reduce the overall cost of the project, standard available wind turbines were

Fig. 3 Bahrain world trade centre-installation of wind turbines on the building. *Source* Internet

used in this project. Expertise drawn from different countries was used in design, manufacturing of plates, erection and commissioning mechanical and electrical system of turbines, integrated with building. Bahrain being an offshore island state in the Gulf region where free, natural wind is available through this project, revolutionary and technological innovation with a new road path has been identified and successfully demonstrated. Other commendable features of this project on wind energy for buildings include the following:

- Potential reduction of carbon footprint
- Heat recovery system
- Windows that can be opened to allow natural ventilating system
- Solar Photo-voltaic-powered outdoor lightning
- Grey water recycling

While BWTC was a pioneering accomplishment in the field of use of wind energy for a building in a large scale utilization through integration of wind turbines with building design, it is reported that "the building was not intended to be a low carbon emission solution" by European and other worldwide standards but in a comparative sense it reduces carbon emissions relative to many other buildings in the Middle East [6]. A recent study by Prof. Blocken, from Netherlands, on this tower with presence of wind turbine to extract wind energy has led to some improvements in the design orientation of the towers with respect to approach wind based on WT measurements and CFD simulation studies, as shown in Fig. 4 [5].

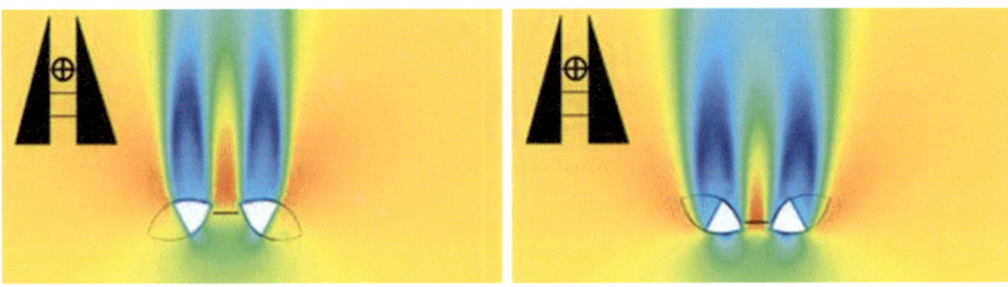

Fig. 4 Flow visualisation using CFD modelling-Bahrain world trade centre [5]

One might also observe that the approach wind conditions for the BWTC corresponded to a coastal terrain with very low level of turbulence intensity. This implies that in general, when we consider such type of projects in a built-up or city environment, performance of such wind turbines will be entirely different which needs to be evaluated based on detailed wind tunnel studies [6].

2.1 Wind Energy Scenario in India

India is ranked 5th position in the world with an installed capacity of 19.9 GW [7]. According to the National Institute of Wind Energy, (known as C-WET earlier) the estimate of official wind power potential is given as equal to 102,778 MW. This corresponds to a hub height of 80 m and with an assumption of 2 % land availability. Quoting a study by Lawrence Berkeley Laboratory, USA, a recent report gives an enhanced India's wind power potential as 2006 GW corresponding to a hub height of 80 m [8]. Tamil Nadu, Andhra Pradesh, Karnataka, Maharashtra, Madhya Pradesh, Gujarat and Rajasthan are the most wind power potential states in India. Wind energy although clean and renewable is often inconsistent in its magnitude and hence poses technical problems of matching the grid frequency and voltage values and thus to network security. To the author's knowledge there are no such tall buildings in India, where use of large wind turbines is used to generate wind induced electricity for making the partial energy consumption demand for the building.

2.2 Small Scale Wind Turbines

Small scale wind turbines, these days are being proposed as an alternate source of wind power as roof top application for buildings in cities or the built up environment. With rotor diameters ranging from 1.2 to 7.6 m are available in the market for producing wind energy. Some of the major disadvantages in case of buildings in built-up environment include:

(a) Reduced wind speed magnitude in the approach wind
(b) Turbulence Intensity levels exceeding above 20 %
(c) Interference effects due to surrounding buildings, trees, plants etc.
(d) The available reduced velocity significantly reduces the amount of wind energy for harvesting.

The turbulence intensity in the wind flow affects the efficiency of the turbine, enhances the noise levels, vibrations which result in reduced lifestyle of the wind energy device. There is a large variability in terms of geometry of the buildings, the pitch roof angle of the buildings, the spacing and configuration of the surrounding buildings in a built-up environment. There are no practical guidelines available for the proper selection of size and position of small scale wind turbines with reference to the building roof which would yield adequate performance with good efficiency. Thus, systematic scientific research investigations using boundary layer wind tunnel experiments are the need of the hour for determining the most suitable locations on the roof top of the building to mount a wind turbine.

The present state of knowledge provides information on wind pressure distribution on roofs of isolated low-rise and high-rise buildings in various terrain conditions, for purpose of load calculations and structural design. However for wind energy extraction purpose, using a wind turbine on such buildings, additional information such as regions of flow separation, shear layer formation, flow reattachment on the roof and velocity and pressure variations across various points on the roof, are also required, whereas available data are far from satisfactory.

2.3 Transportation Sector

Infrastructure development is being focused as one of the important key factors by Government of India in the 12th 5 Year Plan. As a part of it, transportation sector promotes and contributes to national economic development, social integration and security requirements. The total length of National Highways has increased from 22,000 km form year 1951 to about 72,000 km in 2011. The road networks has enhanced from 4 lakh to 42 lakh km during the above period, which is about ten times higher. In India, the National highways are very important, since although they represent only 2 % of total networks, they carry almost 40 % of total road traffic. Hence any efforts towards improving national highways road conditions will significantly enhance efficient freight movement and transport services.

The road mode of transportation in terms of freight, significantly contributes to the total transport sector. For example it is estimated that there will be a great demand of about 68 % for freight through road infrastructure. The fuel consumption in terms of oil and oil products for road sector is about 27 % currently, which is expected to increase to 45 % by 2030 [9]. The total transport sector contributes to about 6.4 % in India's GDP, with road transport alone contributing to 70 %. The amount of crude oil India imported during 2008–09 is about 2.56 millions of barrels per day (mbd), which is expected to increase to about 6.85 mbd in 2030 [2]. With an estimated, more than 300 million vehicles on the road in 2037 in India, the availability and affordability for crude oil, and its emission of Co_2 gas into atmosphere have become grave concerns and challenges demanding engineers and researchers to develop a new innovative technology, using wind as a renewable energy resource.

2.4 Wind Generated Hydrogen as an Energy Resource Carrier

In recent times researchers all over the world have been working on the option of wind-generated-hydrogen as a useful energy carrier derivable from available local wind potential. This would then enable in overcoming the dependence on use of fossil fuels for road transport sector. The requirements of sustainability and eco-friendliness have clearly shown that the burning of fossil fuels mainly causes the emission of green house gases and in this context hydrogen as a transport fuel has received greater attention in several countries including India. For example, the Wind-to-hydrogen (Wind2H2) project carried out by U.S. National Renewable Energy laboratory (NREL, US) along with Xcel Energy in 2009 can be regarded as the pioneering work which demonstrated that hydrogen can be economically produced from natural wind power for transportation fuel purpose [10]. Commercially produced hydrogen has following industrial applications [11].

(i) oil refineries for treatment of crude oil
(ii) food industry
(iii) metal industry and
(iv) fertilizer industry for producing Ammonia

Based on systematic assessment of wind generated hydrogen production potential in Sweden, it is reported that every year about 25,580 ktons of hydrogen can be produced from locally available wind energy which is equivalent to 860 TWh of energy. It was claimed that with only 2 % of above wind generated hydrogen potential in Sweden, the consumption of imported gasoline and the emission of CO_2 can be reduced by 50 % which is very significant [11].

The technology model developed by NREL, USA is shown in Fig. 5. It can be seen that the local wind potential available at the site, can be converted into electricity using wind turbines, and this electricity will be further used for the electrolysis of water whereby water is split into hydrogen and oxygen; the wind generated hydrogen is subsequently compressed, stored, and can be later used to generate electricity through internal combustion engine and supplied to the grid. The author considers the above innovative method developed by NREL, USA as a possible road-path for utilisation of renewable source of wind energy into production of hydrogen which can be subsequently used as a clean, green fuel in the transport sector. It is suggested that this can be adopted in principle, for Indian conditions.

2.5 Challenges and Road Ahead for Growth

Even though the above concept is relatively simpler, its implementation is an uphill task in terms of engineering challenges, which include the following technical challenges between various activities shown in Fig. 5:

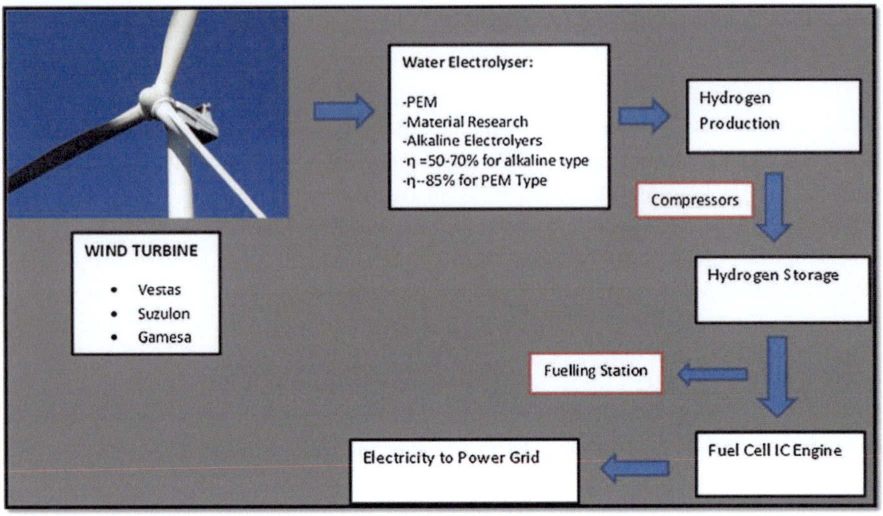

Fig. 5 Wind generated hydrogen technology-flow chart for the method developed by NREL, USA [10]

(a) Selection of suitable site based on wind resource mapping.
(b) Power curve details regarding suitable wind turbines. Presently most of the wind turbines in the Indian market are primarily designed by foreign companies and may not be optimal and appropriate for Indian terrain conditions particularly with respect to turbulence characteristics. This would influence the power characteristics and efficiency of a wind turbine. The annual average wind speeds in many sites in India being about 6–7 m/s, it is reported that the wind turbines presently employed in our wind stations operate at 30–50 % of their rated capacities. This results in the significant reduction of capacity utilization [8].
(c) Structural design capability of supporting towers (say 120 m high), in order to capture high wind power by installing turbines at higher elevations.
(d) Design, installation and commissioning capability of medium scale (500 kW) in commercial scale, for Indian terrain conditions.
(e) Innovation-driven indigenous solutions for integrated power electronics system to link the variable wind output power to the electrolysis technology for splitting water into hydrogen and oxygen.
(f) In a recent study, Shrinet and Govindan [12] from Energy Research Development Association, Vadodara, have demonstrated that wind generated hydrogen can be used as a medium of energy storage and designed a workable model for generating hydrogen from renewable wind using wind turbines and water electrolysis process. The model was designed for application in rural areas with poor grid connectivity, to meet the energy needs of cooking. While

this is a significant achievement, in small-scale, up-scaling for application for the use of hydrogen as a fuel to transport sector poses greater challenges.

(g) Presently in India, there is a heavy shortage in the required design capability and human resources in the areas discussed above [8]. However, efforts are being pursued by several R&D laboratories and industrial organist ions to develop indigenous solutions which would promote advancements in new materials research, reduce considerably import components, enhance in-house manufacturing capabilities with skill development etc.,

(h) The following are some of the organizations, which significantly contribute in the fields of wind energy, wind turbines, water electrolysis, hydrogen storage and fuel cell technology etc., The list is by no means exhaustive.

   i. Hydrogen storage methods: IITs-BHU, Varanasi; Delhi, Mumbai, Chennai; NPL, CSIR, Delhi; CSIR-IICT, Hyderabad
   ii. Wind turbines: CSIR-NAL, Bangaluru, National Institute of wind energy (NIWE), CSIR-SERC, Chennai,
   iii. Fuel cells: CSIR-CEERI, Chennai; BHEL, Hyderabad, TERI, New Delhi, SPIC-Science foundation; Murugappa Chettiar Research Centre, Chennai, ERDA, Vadodara etc.

Thus there is a large scope for cooperative research among various industries, national R&D laboratories and government agencies.

3 Summary and Conclusions

Fossil fuels dominate in meeting the energy demand in all over world including in India. Due to their limited reserves available, added with their characteristics of emitting harmful gases, and since India has a lot of wind potential, use of wind as a promising renewable source of energy is gaining momentum among engineers. Further to wind farm technology using medium scale wind turbines and grid-connected wind power, use of small scale wind turbines (less than 50 kW) as roof-top applications on buildings in the built-up environment are presently receiving increasing interest among the people. However, due to high variability of winds and group/interference effects due to surrounding buildings/obstructions such as sheltering and turbulence etc., systematic scientific investigations are required based on boundary layer wind tunnel studies to formulate guidelines on use of small scale wind turbines to properly make use of wind energy in building sector. Similarly to reduce import of crude oil as fuel for transportation sector and to promote sustainability, an innovative method has been proposed and demonstrated by NREL, USA to use wind generated hydrogen as an alternative source of energy. In the opinion of the author, this may be adopted as a possible road path in India although in many areas such as wind turbines, power electronics for electrolysers, cost effective materials for proton exchange membranes, hydrogen storage etc., focussed and coordinated R&D efforts among academicians, industry and government policies are further required to meet technical and industry needs.

Acknowledgments The author sincerely acknowledges contributions made by various researchers in this field as referred in this paper. Special thanks are to JUET, Guna for their support and motivation.

References

1. Cermak, J.E., Applications of fluid mechanics to wind engineering—a Freeman scholar lecture. J. Fluids Eng. ASME 9–38 (1975)
2. K. Prem, K.V. Raghavan, S.S. Chakraborty, G. Purnendu, Vision, mission and values. INAE-2037, INAE Report (2014)
3. S. Arunachalam, Wind loads on buildings and structures—R & D contributions. Ann. Indian Natl. Acad. Eng. **XII** (2015)
4. www.atkinsglobal.com Bahrain World Trade Centre, Atkins
5. www.tue.nl Bahrain world trade centre is exactly the wrong way around, Prof. Bert Blocken, April (2014)
6. S. Killa, R.F. Smith, Harnessing energy in tall buildings. Bahrain world trade centre and beyond, UAE, Dubai (2008)
7. S. Gomathinayagam, in *Wind power: The way-forward in India*, Proceedings of The Eighth Asia-Pacific Conference on Wind Engineering, December 10–14, (2013), Chennai, India
8. Report on Wind Energy Systems in India-Present status and recommendations for growth. INAE, October (2014)
9. K.P. Singh, Energy scenario in transport sector in India. RITES J. (2009)
10. J. Levene, B. Kroposki, G. Sverdrup, in *Wind energy and production of hydrogen and electricity-opportunities for renewable hydrogen*, Conference paper NREL/CP-560-39534, March 2006
11. H. Shahid, D.M. Siyal, M. Ulla, R.S. Saleem, H. Mark, A preliminary assessment of wind generated hydrogen production potential to reduce the gasoline fuel used in road transport sector of Sweden. Int. J. Hydrogen Energy **40**, 6501–6511 (2015)
12. V. Shrinet, T.P. Govindan, in *Demonstration of sustainable wind hydrogen based distributed energy system*, Electrical Research and Development Association, Vadodara, India, Proceedings of Conference Smartelec (Smart equipment for Smart Grid), paper No. D2-S4, pp. 12–16, (2013)

Waste to Biohydrogen: Addressing Sustainability with Biorefinery

S. Venkata Mohan and Omprakash Sarkar

Abstract Enormous quantity of waste/wastewater is being generated each day from domestic and industrial activities, which comprises a good biodegradable fraction with the inherent potential to generate Biobased products including bioenergy. Mitigating waste generation to permissible limits is an essential pre-requisite to considerably negate its impact on the environment. This communication makes an attempt to delineate the development of a sustainable remediation strategy with dark-fermentative (acidogenesis) process as focal process coupled with value addition via biohydrogen and carboxylic acid production. To facilitate the commercial viability, the process is networked with biorefinery approach to establish waste remediation with resource recovery in the framework of bio-based economy.

Keywords Biohydrogen · Volatile fatty acids · Dark-fermentation · Acidogenesis · Process scale up · Waste biorefinery · Bio-hythane · Bio-plastics · Biodiesel

1 Introduction

Hydrogen gas (H_2) is growing ground as an attractive futuristic energy carrier due to its higher efficiency of conversion, low to non-existent generation of pollutants and high energy density (122 kJ/g; 2.75 folds higher than hydrocarbon fuels). Production of molecular H_2 in industrial scale is mainly from fossil through hydrocarbon reforming. Other processes such as preferential/partial oxidation, stream/auto-thermal reforming, water-gas-shift, desulfurization, plasma/ammonia reforming, thermal decomposition, gasification, etc. were also used. The production of H_2 from fossil fuels is accompanied with the production of greenhouse gases viz., CO_2, CH_4, etc. At present, H_2 production from waste through biological routes

S. Venkata Mohan (✉) · O. Sarkar
Bioengineering and Environmental Sciences (BEES), CSIR-Indian Institute of Chemical Technology (CSIR-IICT), Hyderabad 500007, India
e-mail: vmohan_s@yahoo.com

K.V. Raghavan and P. Ghosh (eds.), *Energy Engineering*,
DOI 10.1007/978-981-10-3102-1_4

utilizing waste as feedstock is emerging as an interesting area of research as it imparts sustainable environment [1–4]. This communication depicts a comprehensive report on the status of current biohydrogen research.

2 Genesis of Biological H_2 Production

In 1939, Hans Gaffron noticed that *Chlamydomonas reinhardtii* a typical green-algae normally grown in pond scum would occasionally change from producing oxygen to hydrogen, what now called as 'Biohydrogen' [5]. Biohydrogen is a natural and transitory by-product of various microbial driven biochemical reactions viz., anaerobic/fermentation/acidogenic, photosynthetic, enzymatic and bio-electrogenic (Fig. 1). Thermo-chemically produced H_2 can also be termed as biohydrogen due to the usage of biomass as feed-stock.

Especially, the passing decade illustrated remarkable research on biohydrogen in both basic and applied fields. Biological H_2 production processes can also be categorized into light-dependent (photosynthesis) and light-independent (fermentation (dark)/acidogenic) processes (Fig. 1). Photosynthetic process can be re-classified into photosynthetic or photo-fermentation process depending on the biocatalyst and carbon source used. Light-dependent processes can be through biophotolysis of water using green algae and cyanobacteria or via photo-fermentation mediated by photosynthetic bacteria. Cyanobacteria and microalgae undergo direct and indirect biophotolysis to produce H_2 by utilizing CO_2 in the presence of

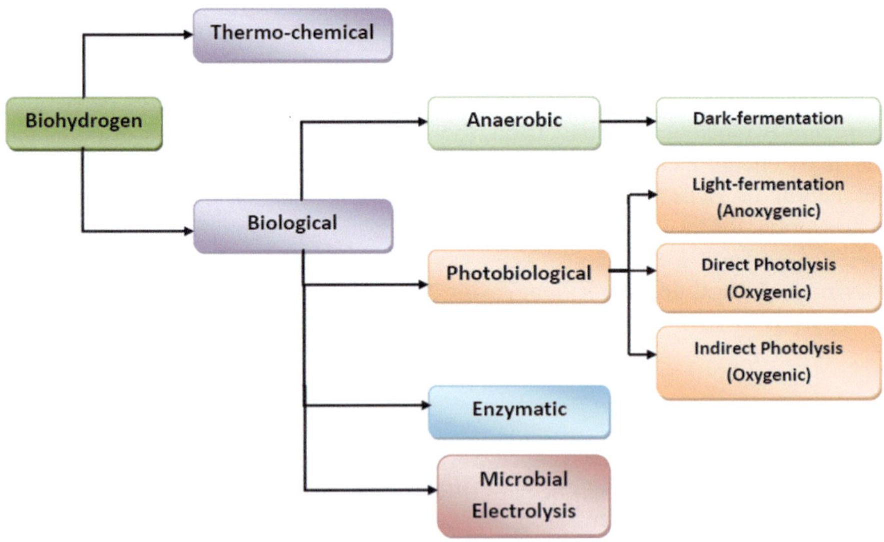

Fig. 1 Illustration of biohydrogen production routes

sunlight and water, while, photosynthetic bacteria manifest H_2 production through photo-fermentation consuming wide variety of substrates ranging from inorganic to organic compounds in the presence of light. Whereas, dark-fermentation process through anaerobic metabolism generated H_2 production is through acetogenic process with simultaneous volatile fatty acids (VFA) and carbon dioxide synthesis by acidogenic bacteria (AB). Both obligate and facultative bacteria are capable of producing H_2 under oxygen free environment. Microbial electrolysis is another route associated with H_2 production in a defined fuel cell operated under applied external potential.

Enzyme mediated in vitro H_2 production, which is one of the fascinating routes envisaged by the scientists. The biochemistry and metabolism involved with the biological routes significantly varies based on the function of biocatalyst used, operating conditions adapted, microenvironment employed and substrate/feedstock used. Initial interest in biohydrogen was visible with the photobiological process. The O_2 sensitivity of H_2 production, combined with the competition between hydrogenases and NADPH-dependent CO_2 fixation are the main limitations.

3 Biohydrogenesis-Dark-Fermentative Acidogenic Process

Fermentative conversion of organic substrate to its end products involves a series of interrelated biochemical reactions viz., hydrolysis, acidogenesis, acetogenesis and methanogenesis (Fig. 2) manifested by five physiologically distinct groups of microorganisms during dark-fermentation process. The complex organic compounds get degraded to form monomers during hydrolysis by hydrolytic microorganisms. Further, these monomers will be fermented by acidogenic bacteria (AB) to H_2 as well as low molecular weight organic acids during acidogenesis. The reversible inter-conversion of acetate production by acetogens and homoacetogens also generates H_2. Finally, the acetoclastic methanogenic bacteria (MB) facilitate conversion of these organic acids and H_2 to CH_4 and CO_2 through methanogenesis. Henceforth, the MB activity needs to be suppressed to make H_2 as a sole metabolic by-product.

4 Waste and Biohydrogen: A Partnership for Sustainable Future

Environmental scientists are shifting focus from 'pollution control' to 'resource exploitation/recovery from waste' to make treatment process sustainable in the context of circular economy [6]. Huge quantities of waste/wastewater are available and are associated with inherent net positive energy which makes them an ideal feedstock. In particular, anaerobic process facilitates the conversion of valued organic waste to useful forms of energy and/or value added products while

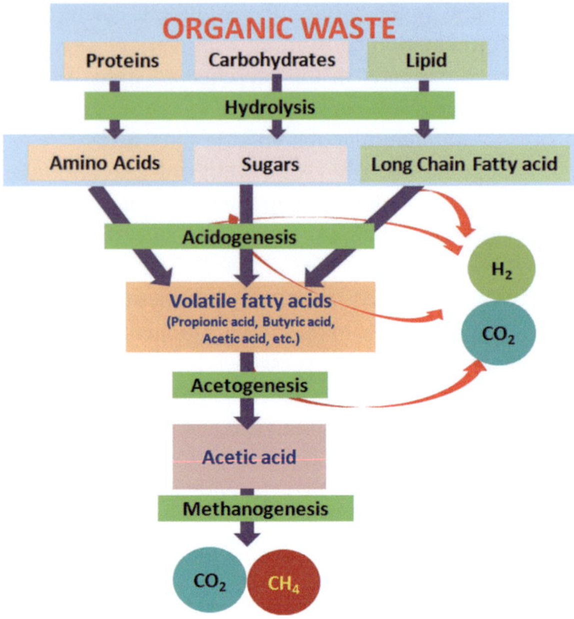

Fig. 2 Schematic illustration of anaerobic conversion of organic substrate through a series of steps

simultaneously achieving pollution control. In conjunction with the wastewater treatment, this process is capable of solving two issues viz., reduction of pollutants in waste and the generation of a clean alternative biofuels. Much of the work reported on biohydrogen production associated to wastewater utility was specifically reported with dark-fermentation process. Dark-fermentative process is relatively less energy intensive and has operational feasibility at ambient temperature and pressure.

5 Biocatalyst

Diverse groups of microorganisms, viz. anaerobic, photosynthetic (heterotrophic and autotrophic) and microalgae are capable of producing H_2 by their specific metabolic routes under pre-defined conditions. Initial stages of research on H_2 were mostly reported with specific strains as biocatalyst in conjugation with defined substrate. After using waste as substrate, attention was focused on using mixed consortia as biocatalyst in the acidogenic process. Mixed cultures offers operational flexibility, diverse functional metabolism, robustness and stability, possibility of using broad range of substrates and moreover avoids the requirement of sterile conditions [7, 8]. From an engineering point of view, producing H_2 by mixed culture offers lower operational cost and ease of control in concurrence to the possibility of using waste as feed-stock [8].

Obligate anaerobes, thermophiles, methanogens and few facultative anaerobes associates with metabolic H_2 production. Evolution of H_2 with anaerobic consortia is often limited as it gets rapidly consumed by methanogens [9–11]. Regulating the metabolic process to acidogenesis and inhibiting methanogenesis helps higher H_2 yields. Pretreatment of biocatalyst facilitates selective enrichment of required consortia specifically acidogenic bacteria (AB) in this case [10, 11]. Optimum growth conditions differences between H_2 producing acidogenic bacteria (AB) and H_2 uptake methanogens bacteria (MB) and forms basis for the pretreatment methods used for the enrichment of H_2 producing inoculum [12, 13]. Pretreatment can also prevents co-existence of other H_2 consuming bacteria due to competitive growth.

6 Factors Influencing Acidogenic H_2 Production

The physiological/physicochemical conditions under which the biocatalyst/ microorganisms produce good H_2 yield are pivotal (Table 1). The function of AB is rate limiting. AB facilitates conversion of carbon under acidic conditions into H_2 and volatile fatty acids (VFA). Under neutral pH the formation of CH_4 is more amenable by MB, while basic pH operation leads to solventogenesis.

AB can function below pH 6.5, while MB operates well between 6.0 and 7.5. The pH range between 5.5 and 6.0 is considered to be ideal to avoid both methanogenesis and solventogenesis [14]. Synthesis of acid metabolites (VFA) causes a drop in the system pH and therefore, reduces the system buffering capacity thereby inhibiting the H_2 production [15]. Maintaining stable redox microenvironment can be possible with good system buffering capacity.

The optimal temperature for H_2 production conjugationally varies with the biocatalyst nature and the type of substrate [8]. H_2 production was reported at ambient, mesophilic, and moderate thermophilic temperatures and a few studies were reported even under extreme thermophilic conditions (over 60 °C) [16, 17].

Table 1 Optimized operational parameters for biohydrogen production

pH	4.5–6.5
Retention time	2–14 h
Temperature	Mesophilic (30–40 °C) and thermophilic (55–70 °C)
Organic load	4–30 g of COD/l
Reactor configuration	Biofilm/suspended
Mode of operation	Batch mode
Substrate	Organic matter with good biodegradability (BOD/COD > 0.5
Light	Not required
Nitrogen	0.10 g/l
Phosphorus	0.60 g/l
Biocatalyst	Selective enriched anaerobic consortia

Both mesophilic and thermophilic temperatures were reported to be favorable for fermentative H_2 production.

Retention time readily manipulates the process and influences the activity of AB as well as MB. Good H_2 yields are reported between 2 and 14 h in various experiments. Longer HRT induces a marked shift to methanogenic from acidogenic process. Reactor configuration and mode of operation influence the performance of an open bioengineering system. Diverse configurations of reactors, viz., biofilm (paked-bed or fixed-bed), fluidized/expanded bed, UASB, suspended growth, membrane bioreactors, immobilized systems, etc., were reported. Biofilm configured systems showed robustness to shock-loads and provide resilience to dynamically changing process dynamics [18]. Batch mode operation in addition to biofilm configuration facilitates dual operational benefits.

7 Waste Biorefinery Approach—Frugal Engineering

Low substrate conversion efficiency due to accumulation of VFA in system and fatty acid-rich wastewater generated from the acidogenic process are major barriers to the practical implementation of this technology. The accumulation of acid by-products causes a sharp drop in the pH resulting in process inhibition. Transforming waste into a usable form by retaining remediation as an integral element enumerates environmental sustainability [6]. Considering the potential of waste/wastewater, research is progressing towards developing a closed loop approach for valorization of waste wherein the effluent coming out of a treatment process is used as a feedstock for another process, thus achieving maximum treatment efficiency [19]. This approach is quite similar to concept of biorefinery, where multifunctional processes are integrated in an optimized sequence to utilize waste with an objective of maximizing the productivity by generation of marketable products (chemicals, materials and bioenergy/biofuels) [15, 20]. I integrating various process in defined and sequential order lead to environmental/waste biorefinery which could be one of futuristic and sustainable solution for bio-based economy. Development of an environmental biorefinery will play a prominent role in maintaining the ecological footprint in the frame work of circular economy.

Environmental and economic concerns suggest using the residual carbon fraction of acidogenic outlet for resource recovery (Fig. 3). Interestingly, biohydrogen can also be viewed as energy source and an key intermediate towards the production of platform chemical (carboxylates/VFA) which can be further transformed as key precursor for bio-plastics (polyhydroxyalkanoates (PHA), [19, 21, 22] lipid (bio-diesel) production by algal heterotrophic cultivation, [23–25] or can be used for hydrogenation of fatty acids to alcohols or can be used as platform chemical [15]. Integrating with the secondary treatment processes viz., methanogenesis, [10] acidogenic fermentation, [26] photo-biological process, [27, 28] microbial electrolysis, [29, 30] etc. were reported. Recently bio-hythane (combination of H_2 and CH_4) production was also reported which ensures fuel efficiency and pollutant

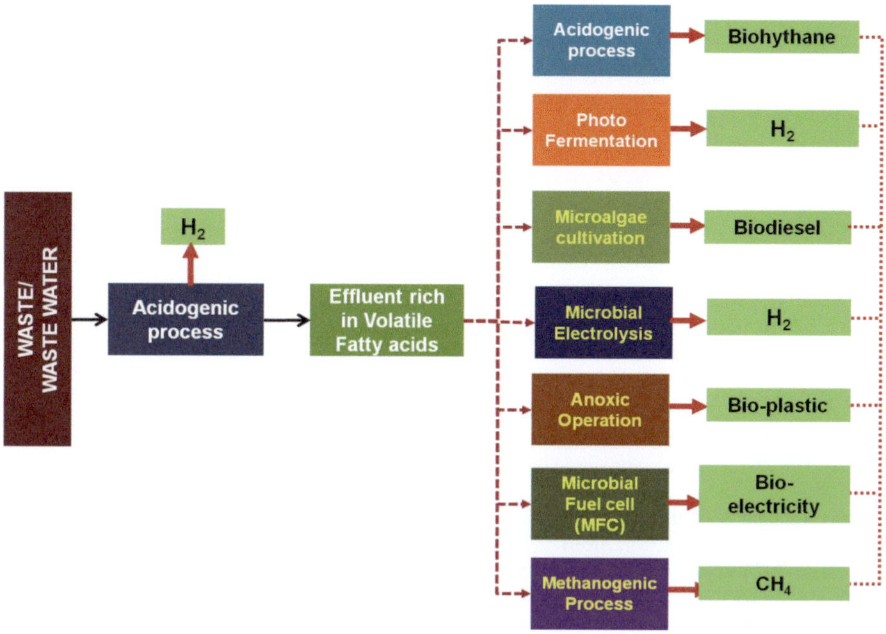

Fig. 3 Schematic representation of various integration processes with dark fermentation (acidogenic process)

reduction [20, 31]. The integration approach address the economic viability and commercialization issues and helps stabilization of the wastewater for environmental restoration [1]. Considering the various possibilities of resource recovery, VFA when separated can be used as platform chemicals viz., acetic acid, butyric acid, propionic acid, etc. with diverse applications in the market [15, 32].

8 Conclusions

Acidogenic process is gaining prominence and importance as a practical and viable method among the biological routes in the context of circular economy and waste biorefinery. This process is relatively less energy intensive, technically simpler, entails low operating costs and is more robust are some of the striking features. To establish an environmentally sustainable biohydrogen technology, optimization of operational factors play important role along with process engineering. In addition, integrated biorefinery approach to overcome some of the persistent limitation by utilizing fatty acids as primary substrate will play crucial role in long run. These integration approaches facilitates reduction in waste load with the advantage of value addition in the form of resource recovery and address circular economy in biorefinery approach. Integration with the existing effluent treatment plant (ETP) is the futuristic goal envisaged with the usage of wastewater as primary feedstock and this which will open up a new prospect for the utilization of renewable and inexhaustible feedstock. With the documented improvements in the performance of this technology over the past 10 years, it can be presumed that a level for practical application can be achieved in a relatively short period of time.

Acknowledgments Authors acknowledge CSIR-Council for Scientific and Industrial Research (SETCA; CSC-0113) and Ministry of New and Renewable Energy (MNRE; No. 103/131/2008-NT) for financial support.

References

1. S. Venkata Mohan, Harnessing of biohydrogen from wastewater treatment using mixed fermentative consortia: process evaluation towards optimization. Int. J. Hydrogen Energy **34**, 7460–7474 (2009)
2. S. Venkata Mohan, Y.V. Bhaskar, T.M. Krishna, N.C. Rao, V.L. Babu, P.N. Sarma, Biohydrogen production from chemical wastewater as substrate by selectively enriched anaerobic mixed consortia: influence of fermentation pH and substrate composition. Int. J. Hydrogen Energy **32**, 2286–2295 (2007)
3. P.C. Hallenbeck, M. Abo-Hashesh, D. Ghosh, Strategies for improving biological hydrogen production. Bioresour. Technol. **110**, 1–9 (2012)
4. S. Venkata Mohan, A. Pandey, *Biohydrogen Production: An Introduction in Biohydrogen* (Elsevier, Burlington, 2013), pp. 1–24
5. H. Gaffron, J. Rubin, Fermentative and photochemical production of hydrogen in algae. J. Gen. Physiol. **26**, 219–240 (1942)
6. S. Venkata Mohan, Reengineering 'Waste remediation' to 'Waste biorefinery' advocating sustainable biotechonomy: new paradigm to frugal innovation. Ann. Indian Natl. Acad. Eng. 300–311 (2016)
7. J. Wang, W. Wan, Factors influencing fermentative hydrogen production: a review. Int. J. Hydrogen Energy **34**, 799–811 (2009)
8. S. Venkata Mohan, Fermentative hydrogen production with simultaneous wastewater treatment: influence of pretreatment and system operating conditions. J. Sci. Ind. Res. **67**, 950–961 (2008)
9. S. Venkata Mohan, G.N. Nikhil, P. Chiranjeevi, C.N. Reddy, M.V. Rohit, A.N. Kumar, O. Sarkar, Waste biorefinery models towards sustainable circular bioeconomy: critical review and future perspectives. Bioresour. Technol. http://dx.doi.org/10.1016/j.biortech.2016.03.130
10. S.V. Mohan, V.L. Babu, P.N. Sarma, Effect of various pretreatment methods on anaerobic mixed microflora to enhance biohydrogen production utilizing dairy wastewater as substrate. Bioresour. Technol. **99**, 59–67 (2008)
11. I. Ntaikou, G. Antonopoulou, G. Lyberatos, Biohydrogen production from biomass and wastes via dark fermentation: a review. Waste. Biomass. Valor **1**, 21–39 (2010)
12. R.K. Goud, S. Venkata Mohan, Acidic and alkaline shock pretreatment to enrich acidogenic biohydrogen producing mixed culture: long term synergetic evaluation of microbial inventory, dehydrogenase activity and bio-electro kinetics. RSC Adv. **2**, 6336–6353 (2012)
13. S. Srikanth, S. Venkata Mohan, L.V. Babu, P.N. Sarma, Metabolic shift and electron discharge pattern of anaerobic consortia as a function of pretreatment method applied during fermentative hydrogen production. Int. J. Hydrogen Energy **35**, 10693–10700 (2010)

14. H. Michael, W. Husemann, E.T. Papoutsakis, Solventogenesis in clostridium acetobutylicum fermentations related to carboxylic acid and proton concentrations. Biotechnol. Bioeng. **32**, 843–852 (1988)
15. O. Sarkar, A.N. Kumar, S. Dahiya, K.V. Krishna, D.K. Yeruva, S. Venkata Mohan, Regulation of acidogenic metabolism towards enhanced short chain fatty acid biosynthesis from waste: metagenomic profiling. RSC Adv. **6**, 18641–18653 (2016)
16. H.S. Shina, J.H. Younb, S.H. Kima, Hydrogen production from food waste in anaerobic mesophilic and thermophilic acidogenesis. Int. J. Hydrogen Energy **29**, 1355–1363 (2004)
17. F.R. Hawkes, I. Hussy, G. Kyazze, R. Dinsdale, D.L. Hawkes, Continuous dark fermentative hydrogen production by mesophilic microflora: principles and progress. Int. J. Hydrogen Energy **32**, 172–184 (2007)
18. P.S. Babu, O. Sarkar, S. Venkata Mohan, Upscaling of biohydrogen production process in semi-pilot scale biofilm reactor: evaluation with food waste at variable organic loads. Int. J. Hydrogen Energy **39**, 7587–7596 (2014)
19. G. Mohanakrishna, S. Venkata Mohan, Multiple process integrations for broad perspective analysis of fermentative H_2 production from wastewater treatment: technical and environmental considerations. Appl. Energy **107**, 244–254 (2013)
20. P.S. Babu, S. Venkata Mohan, Single-stage fermentation process for high-value biohythane production with the treatment of distillery spent-wash. Bioresour. Technol. **189**, 177–189 (2015)
21. M.V. Reddy, S.V. Mohan, Effect of substrate load and nutrients concentration on the polyhydroxyalkanoates (PHA) production using mixed consortia through wastewater treatment. Bioresour. Technol. **114**, 573–582 (2012)
22. M.G.E. Albuquerque, C.A.V. Torres, M.A.M. Reis, Polyhydroxyalkanoate (PHA) production by a mixed microbial culture using sugar molasses: effect of the influent substrate concentration on culture selection. Water Res. **44**, 3419–3433 (2010)
23. S. Venkata Mohan, M.P. Devi, Fatty acid rich effluents from acidogenic biohydrogen reactor as substrates for lipid accumulation in heterotrophic microalgae with simultaneous treatment. Bioresour. Technol. **123**, 471–479 (2012)
24. R. Chandra, S. Arora, M.V. Rohit, S. Venkata Mohan, Lipid metabolism in response to individual short chain fatty acids during mixotrophic mode of microalgal cultivation: influence on biodiesel saturation and protein profile. Bioresour. Technol. **188**, 169–178 (2015)
25. M.J. Griffiths, S.T.L. Harrison, Lipid productivity as a key characteristic for choosing algal species for biodiesel production. J. Appl. Phycol. **21**, 493–507 (2009)
26. M. Gunda, S. Venkata Mohan, P.N. Sarma, Utilizing acid-rich effluents of fermentative hydrogen production process as substrate for harnessing bioelectricity: an integrative approach. Int. J. Hydrogen Energy **35**, 3440–3449 (2010)
27. R. Chandra, G.V. Subhash, S. Venkata Mohan, Mixotrophic operation of photo-bioelectrocatalytic fuel cell under anoxygenic microenvironment enhances the light dependent bioelectrogenic activity. Bioresour. Technol. **109**, 46–56 (2013)
28. T.V. Laurinavichene, B.F. Belokopytov, K.S. Laurinavichius, A.N. Khusnutdinova, M. Seibert, A.A. Tsygankov, Towards the integration of dark- and photo-fermentative waste treatment. Int. J. Hydrogen Energy **37**, 8800–8810 (2012)
29. G.N. Nikhil, G.V. Subhash, Y. Dileep, S.V. Mohan, Synergistic yield of dual energy forms through biocatalyzed electrofermentation of waste: stoichiometric analysis of electron and carbon distribution. Energy **88**, 281–291 (2015)
30. D. Call, B.E. Logan, Hydrogen production in a single chamber microbial electrolysis cell lacking a membrane. Environ. Sci. Technol. **42**, 3401–3406 (2008)
31. J.Y. Lee, J. Yun, T.G. Kim, D. Wee, K.S. Cho, Two-stage biogas production by co-digesting molasses wastewater and sewage sludge. Bioprocess Biosyst. Eng. **37**, 2401–2413 (2014)
32. F.C. Silva, L.S. Serafim, H. Nadais, L. Arroja, I. Capela, Acidogenic fermentation towards valorisation of organic waste streams into volatile fatty acids. Chem. Biochem. Eng. **27**, 467–476 (2013)

Indian Advances in Fast Breeder Nuclear Reactor Engineering

Baldev Raj and P. Chellapandi

Keywords FBR · Nuclear reactor design · Nuclear reactor prototype · India's nuclear energy program · Nuclear reactor construction

1 Introduction

Growing concern regarding environmental issues; local pollutions and climate changes threatening sustainability nuclear energy is a preferred option for base load electricity in India. Nuclear energy can provide sustainable energy over centuries with the available uranium and thorium resources. A special mention needs to be made for Fast Breeder Reactor (FBR) with closed fuel cycle as an efficient sustainable system for the effective utilization of uranium (>90 %) with significantly reduced high level radioactive waste and of less toxicity to be managed. FBR can also be designed to incinerate high level nuclear wastes arising from the reprocessing of spent fuel from thermal nuclear reactors. The advantages of FBR are illustrated in Fig. 1.

In the Indian context, the available uranium can feed 275 GWe for at least 200 years, through FBR and thorium can be utilized through both fast and thermal spectrum cycles beyond 2050 (Fig. 2). In view of its importance for the long term energy security, India started the FBR programme as early as in 1972, by constructing a sodium cooled loop type 40 MWt/13.2 MWe Fast Breeder Test Reactor (FBTR) at Kalpakkam. FBTR was commissioned in 1985 with unique plutonium rich carbide fuel. The fuel achieved record a burn-up of 155 GWd/t in MARK-I core without even a single clad failure. Further, the reactor life has been extended by 20 years beyond the first 25 years and thus continue serving as an irradiation facility for the fuels and core structural materials. Based on the experiences gained

B. Raj (✉)
National Institute of Advanced Studies, Bangalore 560012, India
e-mail: baldev.dr@gmail.com

P. Chellapandi
Bharatiya Nabhikiya Vidyut Nigam Limited, Kalpakkam 603102, India

K.V. Raghavan and P. Ghosh (eds.), *Energy Engineering*,
DOI 10.1007/978-981-10-3102-1_5

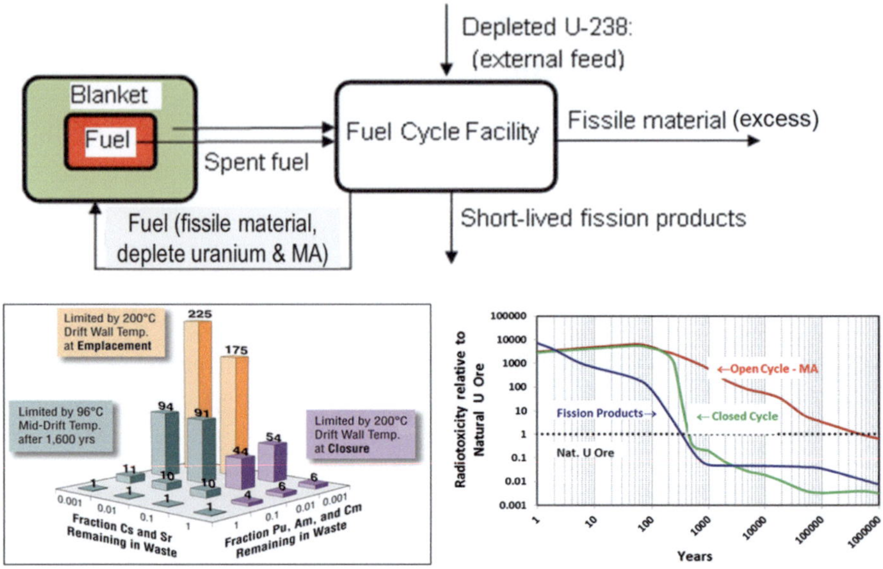

Fig. 1 Advantages of FBR with closed fuel cycle

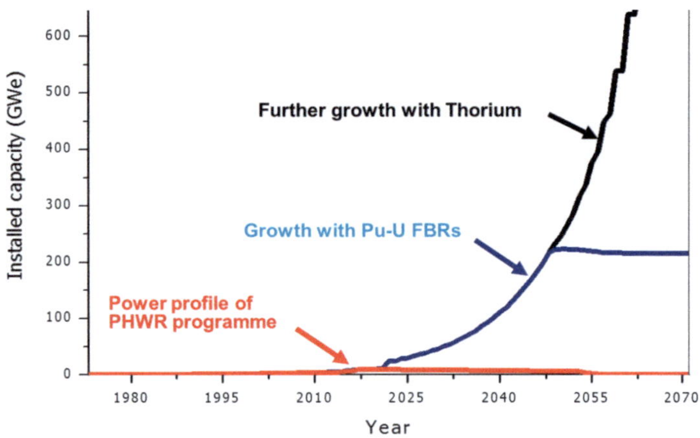

Fig. 2 Strategy for energy growth rate in India

with international FBRs including FBTR, a 500 MWe capacity Prototype Fast Breeder reactor (PFBR) was conceived in 1985 by Indira Gandhi Centre for Atomic Research for demonstrating the techno-economic viability of a series of large size FBRs planned by the Department of Atomic Energy. The construction of PFBR was started in 2004 by BHAratiya Nabhikiya VIdyut Nigam (BHAVINI). Currently the reactor is under commissioning phase and the first criticality is envisaged in

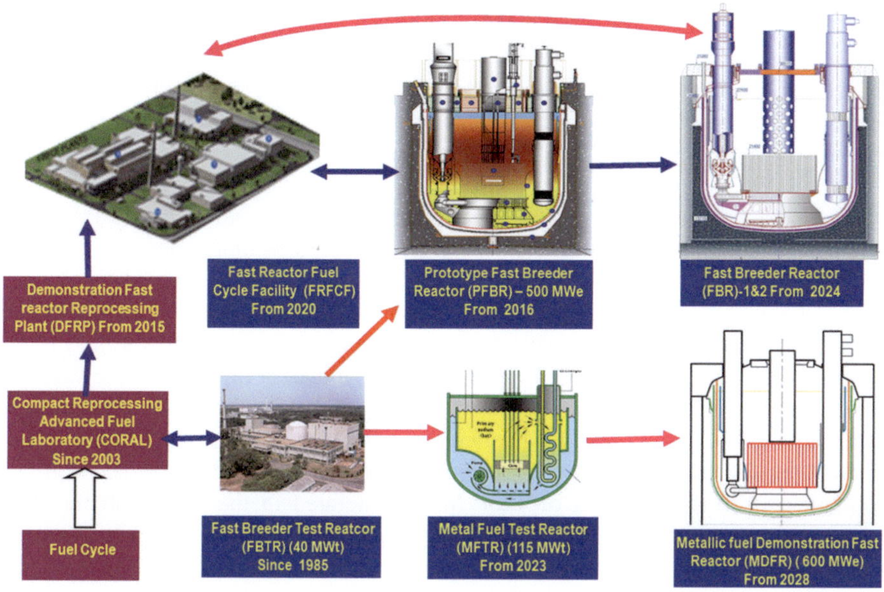

Fig. 3 Fast reactor and associated fuel cycle programme in India (up to 2030)

2016. The power operation will start within a month followed by nominal power generation. Beyond PFBR, it is planned to start construction of two FBRs (FBR-1&2) incorporating several advanced features to provide electricity at a competitive cost with enhanced safety features on par with internationally evolving safety criteria. Enhancing the power to 600 MWe capacity from existing 500 MWe with introduction of innovations in technology. Twin unit concept sharing several common facilities particularly, fuel building, decontamination facility and maintenance building, longer design life of 60 years in comparison to current 40 years with higher load factor of 85–90 % and high burnup (150 GWd/t) are a few notable improvements as compared to PFBR. Towards achieving rapid growth rate, metallic fuel based reactors are also planned beyond 2030 after deriving adequate confidence through construction and operation of Metallic Demonstration Fast Reactor (MDFR) by IGCAR. The roadmap conceived up 2030 is illustrated in Fig. 3. The confidence on the success of FBR programme is based on our sound knowledge, sustained in-house R&D with national and international collaborations, technological achievements and construction & operation experience accumulated over more than 40 years.

In this short article, we focus on the challenging aspects of reactor engineering of Sodium cooled Fast Reactor (SFR), achievements made in the design, construction and commissioning of PFBR. Specific aspects covered are in the domain of metallurgy of core and coolant circuit materials and technologies related to manufacturing and erection. Technical details of PFBR are provided to start with.

2 Technical Details of PFBR

PFBR is a pool type SFR and uses Mixed Oxide (MOX) fuel with ~25 % plutonium oxide and 75 % depleted uranium oxide. The PFBR is constituted by reactor assembly and three major heat transport circuits as shown in Fig. 4. The primary sodium removes the heat while passing through the core and mixes with the hot pool (primary circuit). The sodium from the hot pool in turn transfers its heat to the secondary sodium flowing in the Intermediate Heat Exchangers (IHX) and joins to the cold pool. The flow of sodium from the cold pool to the hot pool is facilitated by Primary Sodium Pumps (PSP). There are four IHX and two PSPs, housed within main vessel along with core and sodium. In the secondary loop, the secondary sodium, coming out of IHX transfers the heat to the water in steam generators (SG) to generate steam and flows back to the IHX. There are two secondary loops with four SGs per loop. The steam generated at ≅490° C and 17 MPa is fed to a convectional steam water system to generate electricity of 500 MWe from the net heat of 1250 MWt generated in the core with the thermodynamic efficiency of ≅40 %. The design of PFBR envisages a replacement of 1/3rd spent fuel subassemblies (66 nos.) in every 8 months with the fresh subassemblies after 6 months of nominal power operation, implying the load factor of 75 %. The residential period of the fuel subassemblies in the active core is 8 months and 16 months in internal storage space in the core for the equilibrium cycle.

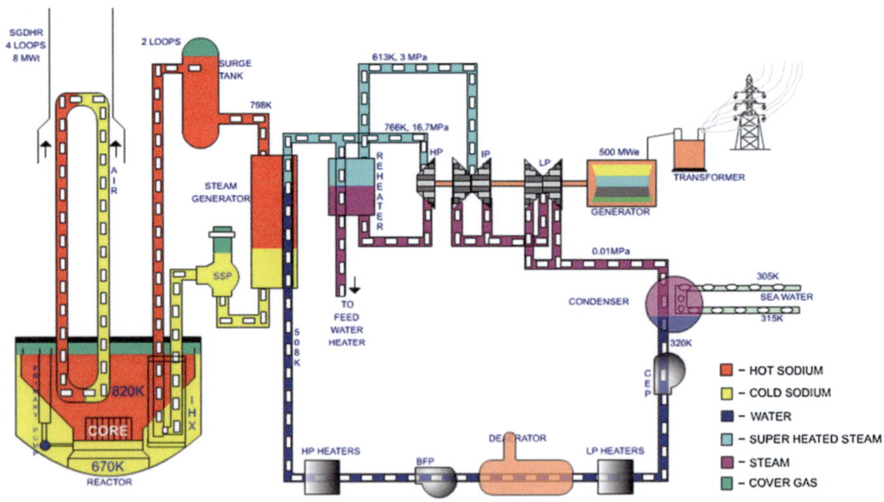

Fig. 4 Heat transport circuit of typical SFR

3 Reactor Engineering Aspects of SFR

Reactor engineering starts with process design followed by mechanical design. While the process design provides optimum dimensions of components such as diameter, height, length, etc., the mechanical design recommends a set of parameters such as structural wall thickness, manufacturing specifications etc. for ensuring the structural integrity over the specified design life of the components. Accordingly, the design involves choice of optimum process parameters, appropriate high quality materials, applicable design criteria, choice of codes and standards, robust construction methodologies and inspection strategies. Design also addresses the importance of the components that quantifies the consequence of the failure of the component w.r.t safety. Accordingly, the components are classified as safety class-1 and safety class-2 in case of nuclear plants. Figure 5 depicts the choice of materials and safety classification of reactor assembly components for the PFBR. The design should incorporate adequate factors of safety on structural wall thickness for withstanding the mechanical and seismic loadings, which however need confirmation through detailed mechanical analysis with validated computer

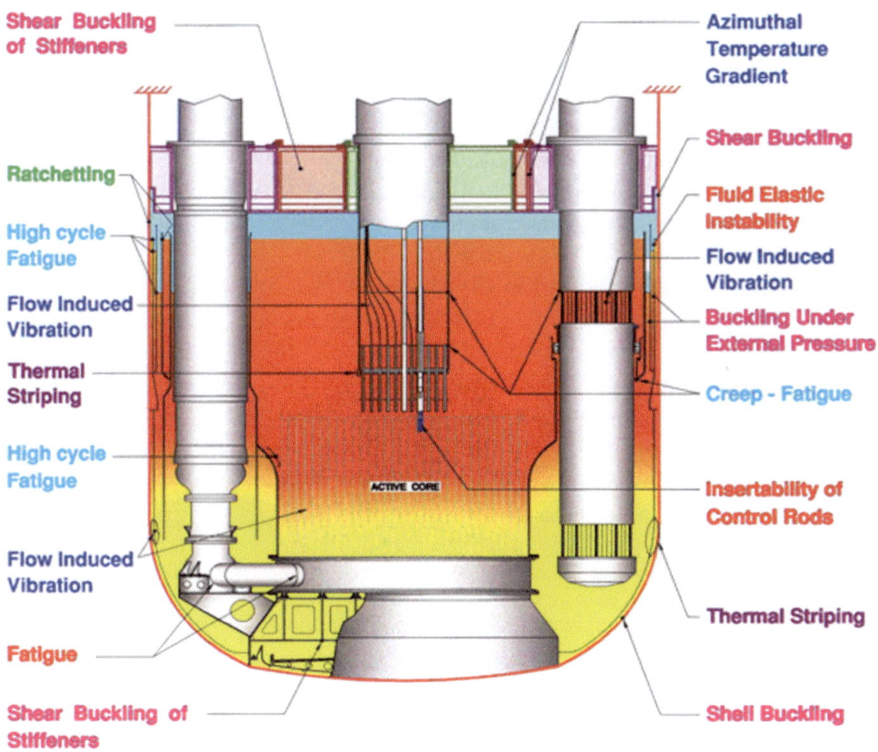

Fig. 5 Failure modes for structural design of SFR reactor assembly

codes, experimental testing in simulated environment representing sodium purity, temperature and stresses in steady, transient and accidental conditions. Robust safety criteria are formulated for ensuring safety, i.e. for preventing and or mitigating radioactivity release in case of accidents, by categorizing the transients under two broad categories viz. design basis and beyond design basis events depending upon the probability of occurrences. Such safety design demands for incorporation of provisions, such as diversity in shutdown and decay heat removal systems for the design basis events, so as to prevent beyond design basis accidents. These apart, defense-in-depth strategy is adopted through which novel design features such as re-criticality free core, effective core catcher, and containment concepts are introduced for the severe accident management. Critical examination and consideration of the feedback experience are essential to demonstrate robustness.

A few of the above aspects are elaborated in the following paragraphs in the context of a Sodium Fast Reactor (SFR).

4 Material Choice and Issue

The choice of structural materials broadly depends upon the fuel (metallic or ceramic) for the core and on the coolant (gas, sodium, lead) for the heat transport circuit components. The structural materials, especially for the core should respect several requirements with the presence of neutron irradiation effect. **A** few major requirements are **given** in the Table 1.

The materials for the out-of-core in the reactor assembly face less severe environment in terms of temperature and neutron dose (thanks to heavy shielding around the fuel and blanket zones), but their design life is much longer (~60 years). The materials chosen for PFBR are listed in Table 2.

Challenging aspects involved in the choice of materials is demonstrated from two case studies: (1) SG material and (2) hard facing material chosen for all nuclear steam supply system components of PFBR made of 316L(N) and 304L(N) SS.

Modified 9Cr1Mo is a ferritic steel, chosen for SG. The weld joints of this material experience premature failure in the heat affected zones (HAZ), in the sense

Table 1 Major requirements for core structural materials

Sl. No.	Component	Major requirements
1	Clad	Swelling resistance, creep strength, high mechanical strength, adequate ductility, compatibility with fuel, fission products and coolant. Fabrication and weldability with current technologies
2	Wrapper	Swelling resistance, creep strength, moderate mechanical strength, adequate ductility, compatibility with coolant. Fabrication with current technologies
3	Spacer wire	Swelling resistance, moderate mechanical strength, compatibility with coolant. Fabrication with current technologies

Table 2 Material choice for PFBR

Sl. No.	Components	Materials
1	Clad and Wrapper Tubes	20 % cold-worked D9 steel
2	Grid plate	SS 316L(N) (Though temp <700 K)
3	Steam Generator (SG)	Ferritic steel (Modified 9Cr-1Mo)
4	Top Shield (Roof-slab)	Spl. Grade Low carbon A48P2 steel
5	Other structural components	Low carbon Austenitic SS alloyed with nitrogen up to 0.08 wt% (LN types) SS 304LN for <700 K SS 316LN for >700 K

that creep rupture life of weld joints are less compared to the base metal. The failure location changes from base metal to inter-critical region of HAZ with increase in test temperature and decrease in applied stress. This kind of failure that occurs at the inter-critical HAZ as a result of preferential accumulation of creep strain with associated creep cavitation and poor creep strength is called as Type IV cracking. R&D studies carried out by IGCAR and elsewhere have enabled the development of suitable compositions, welding geometries and parameters for modified 9Cr-1Mo steel to improve, Type-IV cracking resistance.

Regarding hard facing material, a nickel-base cobalt-free ERNiCr-B is chosen to replace the cobalt-base alloys for transfer arm and grid plate. Components plasma transferred-arc welding (PTAW) process is adopted instead of the conventional GTAW process. The PTAW deposition procedures have minimized dilution of the coating by the austenitic stainless steel substrate and enabling achieving improved mechanical properties thereby ensuring adequate performance of the component.

5 Reactor Design: Challenges and Achievements

Design should address all the possible failure modes comprehensively, giving due considerations to the operating experiences, which have accumulated about 400 reactor years. High temperature design for long reliable operation of components operating at temperatures around 550 °C for design life of 40 years, design of fuel handling machines, shutdown systems mechanisms and rotating equipment operating in sodium and argon cover gas space, provisions for managing sodium fire followed by large sodium leaks and sodium water reactions in the steam generators, seismic analysis of interconnected buildings resting on the common base raft, seismic design of thin walled vessels, pumps and absorber rod mechanisms and in-service inspection of reactor internals within sodium at higher temperatures are a few challenging issues which have been addressed in the design.

SFR design in general provides excellent opportunities to pursue R&D in many frontline areas such as thermal hydraulics, structural mechanics focusing on flow instabilities, gas entrainments, thermal striping, stratifications, ratcheting demand high quality research expertise. Figure 5 depicts failure modes that are considered in the structural design of SFR.

6 Technological Challenges

6.1 *Specific Features of SFR Components W.R.T Manufacturing and Erection*

SFR components, in general are characterized as large diameter thin walled shell and slender structures. Tight manufacturing tolerances are specified to enhance their buckling strength as well as to have minimum vessel dimensions. In the reactor assembly, main vessel, thermal baffles, inner vessel, core support structure and grid plate are to be positioned sequentially maintaining the coaxiality with the safety vessel, so that the core central line is in-line with the central lines of coaxial vessels: one of the requirements to facilitate smooth operation of control rods as well as for facilitating accurate monitoring of temperature of sodium emerging from the core subassemblies. Further, they have to be erected accurately to maintain the annular gaps for the uniform sodium flows and temperatures. During manufacturing stage, single side welds are unavoidable at some difficult locations particularly in the case of box type structures. In-service inspection is not possible for all the components in sodium and hence, stringent quality control is required in the pre-service level itself. From the dimensional stability point of view, residual stresses should be kept to minimum value by adopting robust heat treatment processes and mockup trials. It is preferable to use minimum number of materials from the consideration of economy and material data generations. This approach enhances reliability of performance of materials in the operation. Austenitic stainless steels, the main structural material in particular, call for careful considerations for welding without significant weld repairs and distortions. Construction experience of international and PFBR indicates that reactor assembly components decide the project time schedule, even though their cost is relatively small compared to civil, structures sodium circuits and Balance of Plant (BoP). Only limited experience on manufacturing and erection of components exists internationally for the application. These apart, the design and manufacturing codes are evolving. These are the major challenges in the manufacturing and erection of reactor assembly components.

6.2 Manufacturing of Thin Shell Structures

Reactor assembly components of SFR are characterized by large diameter thin walled shell structures made of austenitic stainless steel Type 316 LN. They are fabricated as welded assembly of large dimensioned petals. High quality welds achieved with insignificant weld repairs, cold forming limiting to <10 % and close form tolerances of ±12 mm (≤ 0.2 % R) are the major achievements. These are accomplished by adopting stringent dimensional control on petal level and subsequently robust weld fit up and weld sequence methodology, state-of-art techniques for inspection and quality control, numerical simulations of forming and welding procedures, innovative mockup trials, lessons learnt from feedback experiences of various industries and elaborate technology development exercises.

6.3 Manufacture of Large Diameter Grid Plate

Nozzle welding with least distortion, large diameter colmonoy hard facing without defects, hard facing of large number of sleeves on the inner diameter at a depth of 500 mm are some of the achievements accomplished during the manufacturing of grid plate. These are achieved by adopting state-of-art methods in dimensional control, innovative distortion control methods, modern dimensional inspection methods, handling of plates with large percentage of perforation, innovative way of heat treatment of large dimensioned components and novel techniques indigenously developed for handling band assembly of large number of items (>14,000).

6.4 Manufacturing of Steam Generators (SG)

SG is made of modified 9 Cr 1 Mo steel. It is **a** tall component of about 25 m height and has more than 500 tubes. The most challenging task is in-bore welding of tube-to-tube sheet joints, which are carried out with stringent acceptance standards on dimensions and weld quality without any lack of penetration & fusion, cracks, undercuts and unacceptable porosity. The maximum concavity achieved is practically zero and maximum weld thinning is less than the permissible <0.2 mm. Welding technology has been matured based on elaborate technology development exercises and many trials.

6.5 In-Service Inspection (ISI) of Main Vessel

The main vessel is guarded against any sodium leakage by the outer safety vessel, which is placed concentric to the main vessel. The continuous surveillance of the integrity of the vessel is supplemented by periodic examination of the main reactor vessel, safety vessel and core support structure, performed when the reactor is shut down using remote controlled robotic vehicle. The ISI involves visual inspection of surfaces and volumetric inspection of welds on the main and safety vessels using ultrasonic and eddy current techniques during fuel handling operations. The comprehensive system consists of independent devices which can move inside the interspace between main and safety vessels carrying inspection modules for eddy current, ultrasonic and visual inspections. Each device is provided with navigation camera modules for guiding the devices in the interspace during ISI. The devices are provided with essential sensors such as resolvers, temperature sensors, inclinometers, LVDT, load cells, etc., for monitoring and control of the robotic vehicles during ISI. The components are sophisticated and inspections have to be done within very narrow spaces. The area to be scanned is very large which needs to be completed within possibly shortest time (has bearing on the load factor of the plant). These robotic devices are fully computer controlled for remote operations developed with intelligent algorithms. Extensively validated inspections based on numerical simulations and extensive manufacturing technology development exercises. Indigenous efforts addressing all the challenges are central to these achievements.

6.6 Lessons Learned from Manufacture of PFBR Components

Technology development prior to start of construction is essential for long delivery components. Judicious choice of tolerances, number and location of welds and appropriate and adequate inspections has to be made. Robust criteria need to be applied for the acceptance of manufacturing deviations and material compositions. Indigenous materials have been used after qualifications of manufacturing processes with a large number of innovations along with best of practices and standards. Industry needing technical support, desires decisions to be taken expeditiously basis of science and experience. Our considerations incorporate best of our international experiences.

7 Current Status of PFBR Project

Construction of PFBR is completed. The most challenging aspect has been accurate erection of thin walled reactor assembly components, which has been successfully completed with excellent coordination of IGCAR (organization responsible for design), BHAVINI (organization responsible for commissioning and operation) and manufacturing industries (Fig. 6). Currently, stage-wise commissioning is in

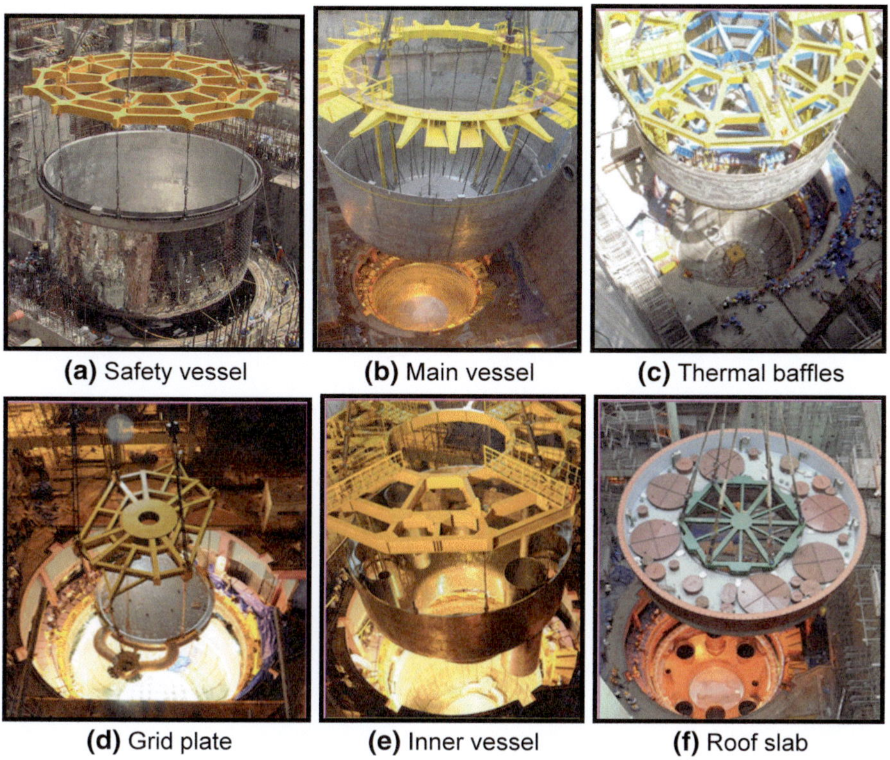

Fig. 6 Installation of reactor assembly components of PFBR

progress starting from sodium circuits. After completing this, fuel loading shall start by replacing the dummy core subassemblies with blanket and fuel subassemblies. Fuel subassemblies will be loaded batch by batch to check for critical configuration. Once neutron flux level is sustained, reactor is said to become 'critical'. This phase is expected to reach in 2016. After completing the specified reactor physics tests, reactor power will be raised with generation of steam in steam generators. With steam pumping into turbo-generator, electricity will be generated and the commercial operation shall be achieved.

Acknowledgments Eminent and wonderful colleagues from IGCAR, BHAVINI, DAE family, collaborating academicians, researchers, industrialists in the country and the world are saluted and worthy of our gratitude and sincere mention.

China's Long Road to the High-Efficiency, Clean and Low-Carbon Energy Transition

Suping Peng

Abstract China, the largest energy producer and consumer in the world, has developed a comprehensive energy supply system consisting of coal, crude oil, natural gas, nuclear energy and renewable energy resources. The country has to strike a balance between satisfying its huge energy demand and safeguarding the environment in order to make its economic growth sustainable. To build a secure, economical and clean energy industry, China relies on scientific, technological and system innovation. This paper aims to introduce the current conditions and challenges in China's energy development, its energy policies and energy revolution strategies, comprehensive measures and solutions for the energy transition, and the high-efficiency, clean and low-carbon coal utilization. China is exploring and practicing a new way in the high-efficiency, clean and low-carbon energy transition to ensure its sustainable energy development, which will offer a wide range of co-benefits in the future.

Keywords Energy development · Low-carbon energy transition · Coal utilization · China

1 Introduction

Along with the rapid development of China's economy and its people's livelihood level, energy remains a major strategic issue [1], though the country has achieved great processes in its total energy production and consumption, as shown in Fig. 1. In 2014, the total primary energy output of China amounted to 3.6 billion tce [2], the world's largest energy supply system, of which the output of raw coal reached 3.87 billion tons, crude oil 0.21 billion tons, and natural gas 130.2 billion cu m. In

S. Peng (✉)
Chinese Academy of Engineering, Beijing 100088, China
e-mail: psp@cumtb.edu.cn

S. Peng
China University of Mining and Technology (Beijing), Beijing 100083, China

K.V. Raghavan and P. Ghosh (eds.), *Energy Engineering*,
DOI 10.1007/978-981-10-3102-1_6

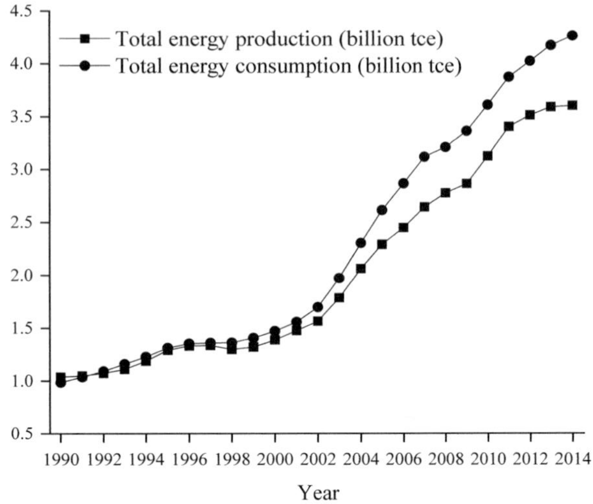

Fig. 1 Primary energy production and consumption in China, 1990–2014

this year, the electricity production amounted to 5.6 trillion kWh, and thermal power accounted for 74.9 % of the total. Meanwhile, the total primary energy consumption in 2014 amounted to 4.26 billon tce, and non-fossil energy contributed 11.2 % to the national total [2]. The installed generating capacities of hydropower and wind power are all rank first in the world.

Despite these promising trends of energy development, tremendous challenges still lie ahead. China has a much higher proportion of coal in energy sources, and also of coal in direct fuel use and electricity/heat generation, while globally energy supply is more dominated by oil and gas [3]. Meanwhile, the energy demand is more dominated by investment-driven industrial production such as steel, chemical and cement production and also producing goods/services for export [3, 4]. On the whole, the energy challenges of China include "*prominent resources restraint, low energy efficiency, increasing environmental pressures, grave challenges to energy security and reforms called for current systems and mechanisms*" [1]. Prominently, the environmental issue such as greenhouse gas emissions associated with energy production and utilization has become a pronounced global concern [5]. For instance, the energy-related CO_2 emission in China 2014 was 9761.1 million tons, about 27.5 % of the world total [6]. Aside from concerns about climate change, air pollution is already affecting almost all major energy-consuming regions with the poor air quality in the urban centers. The main sources of CO_2 emissions and air pollutants such as SO_2, NO_x and particulate matters can be attributed to the fossil fuel consumption such as coal combustion and burning transportation fuels [5]. In addition, the water crisis and ecological damage associated with large-scale energy exploitation and utilization are emerging in the central and western provinces, where untapped coal and other energy resource reserves are abundant.

In fact, the level of per-capita energy consumption is still low in China. The country's total energy consumption will continue to rise along with the urbanization and industrialization progress. Particularly, fossil fuels will still dominate the energy consumption mix for a long time to come, which poses a growing challenge in the environment protection and addressing climate change. Thereafter, the energy development of China must go through a route characterized by the high-efficiency, clean and low-carbon energy transition.

2 China's Energy Policies and Energy Revolution Strategies

China's governments at all levels have taken measures to maintain sufficient and efficient supply of energy resources, control vastly growing energy consumption and deal with negative environmental effects. The basic energy policy frameworks of China have been addressed in the whitepaper on *China's energy policy 2012*, which include "*giving priority to conservation, relying on domestic resources, encouraging diverse development, protecting the environment, promoting scientific and technological innovation, deepening reform, expanding international cooperation, and improving the people's livelihood*" [1].

In practice, China plans to launch a national emission trading system by 2017. In the *Energy Development Strategy Action Plan of China* (2014–2020), the national total energy consumption by 2020 will be kept below 4.8 billion tce and its coal consumption 4.2 billion tons around. In addition, the installed generating capacities of hydro-power, wind power, solar power and nuclear power will increase to 350, 200, 100 and 58 GWe, respectively [7]. Specifically, the Chinese government has made the commitment that its CO_2 emissions per unit of gross domestic product (GDP) by 2020 will be 40–45 % lower than in 2005, and 60–65 % lower by 2030. Meanwhile, the share of non-fossil energy in its total primary energy consumption will increase to 15 % by 2020 and 20 % by 2030, and its carbon emissions will peak around 2030.

In 2014, Chinese President Xi Jinping urged more efforts to revolutionize the country's energy sector and further ensure national energy security. The energy revolution strategies of China cover the five parts of energy consumption reform, energy supply reform, energy technology innovation, reforms in pricing mechanism and the deepening of international cooperation. Energy consumption reform mainly refers to control the total energy consumption and rein in irrational energy consumption. Energy supply reform refers to establish a diversified energy system and achieve cleaner coal use. The principal objective in energy technology innovation is to fuel economic growth. The main content of reforms in pricing mechanism contains the development of a competitive energy market. If these ambitious energy revolution strategies can be achieved, the energy trajectory of the world's largest energy consumer will be fundamentally shifted.

3 Comprehensive Measures and Solutions for the Energy Transition

One of the most effective ways for the energy transition in China is the high-efficiency and clean utilization of traditional fossil fuels. For the coal production and utilization, China has enhanced production capacities and mechanization level in coal mining, and promoted clean and highly efficient development of thermal power such as supercritical and ultra-supercritical coal plants. It also develops some thermoelectricity co-generation units and the integrated coal gasification combined cycle (IGCC) power plant [1]. In the past decade, several demonstration projects of carbon capture, utilization and storage (CCUS) have been completed and operated. As to conventional oil and gas resources, China steadily increases the reserves and outputs of crude oil in major oil-production regions [1]. The country promotes the productivity of natural gas, increases the gas production in major gas fields, and pushes forward the offshore oil-gas development. China also enhances the exploration and exploitation for unconventional gas. The proven geological reserves of coal-bed gas and shale gas have increased rapidly in recent years.

Aside from concerns about the fossil fuel utilization, China actively raises the proportion of non-fossil energy in the primary energy mix. Developing nuclear energy and renewable energy has become the main measure to optimize the energy structure and foster emerging energy industries. The development of nuclear energy in a safe and highly efficient way is of great significance for promoting the clean energy development on the east coast. Currently, the feasibility studies have been announced for the construction of inland nuclear plants. China develops hydro-power actively and has speeded up the construction of hydro-power stations at different scales and pumped-storage power stations [1]. China also promotes diverse patterns of wind power and solar power development in terms of intensive exploitation and distributed utilization. As to the improvement of power grids' renewable power integration ability, it is speeding up grid construction and improving the performance of power equipment. In addition, China expedites the development of bio-fuel, biogas, geothermal energy, tidal energy and other renewable energy resources [1].

Engineering measures and solutions are generally the key for the energy transition. a series of advanced energy technologies have been developed and deployed in China, such as the high-efficiency and intensive mining technologies, direct coal liquefaction and coal-to-olefins technologies, high-efficiency clean coal power generating technologies, offshore wind power technologies, solar thermal power technologies, and extra-high-voltage power transmission technologies. Meanwhile, the technological advance of key equipment has been boosted in recent years, like high-capacity and high-parameter generating units and third-generation nuclear power plants, and other advanced power generation systems.

4 High-Efficiency, Clean and Low-Carbon Coal Utilization

Since coal is the dominant energy resource in China (around 70 % in its total energy mix), as shown in Fig. 2, the fundamental position of coal in the country's primary energy mix will be long-term [8]. Coal combustion is also the major source of air pollution and greenhouse gas emissions [9, 10]. A significant decrease in coal demand by the adjustment of the energy mix and industrial structure requires long time frames and considerable technological improvements. Historically, coal consumption has a high correlation with the GDP growth in China, and the impact of direct reduction in coal consumption can be negative on economic growth [11, 12]. Therefore, the energy policies or strategies aiming to guide low-carbon economy development in China rely highly on the high-efficiency, clean and low-carbon coal utilization [5, 13].

Developing and deploying clean coal technology is a continuing task that can offer long-term global environmental and climate benefits. China is well advanced in its development of high-efficiency coal power plants with the large capacity and high parameters. Supercritical and ultra-supercritical power generation stations have been installed rapidly. An example is Shanghai Waigaoqiao No. 3 power plant, one of the best in the world, which has an average coal consumption of 276 g/kWh for power generation at 75–81 % capacity, much lower than the current average level of China's coal power plants, 319 g/kWh in 2014 [5].

Among various kinds of clean and efficient coal utilization methods, one of the most valuable options is coal gasification to produce chemicals, fuels and electricity [12]. Generally, the efficiency is approximately 40 % for the conventional coal power generation. It has been proved that this efficiency can be increased by the IGCC plants, but the IGFC (integrated coal gasification fuel cell combined cycle) system has the potential to achieve higher efficiency than IGCC. The core technologies of IGFC include SOFC Power Generation Technology, CO_2 electrolysis

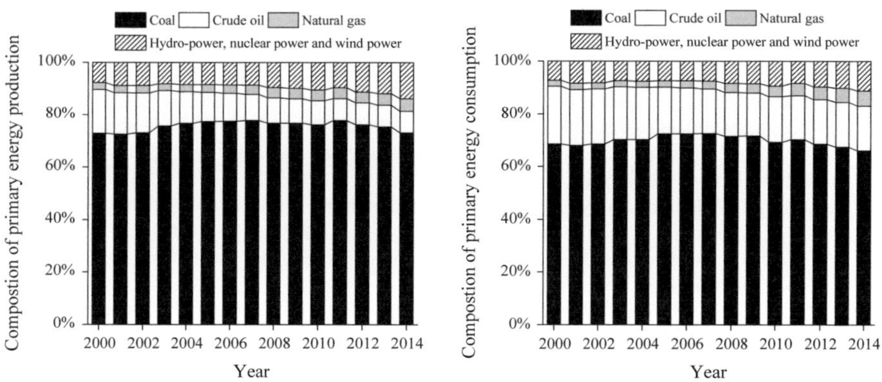

Fig. 2 Compositions of primary energy production and consumption in China, 2000–2014 [2]

technology (SOEC), CO_2 separation membrane technology, and oxygen transport membrane technology. This system is characterized by high power generation efficiency, fuel flexibility, low economic cost, efficient water cycle, and prominent CO_2 capture capacity. For instance, the operating temperature of coal gasification-driven SOFC is high and the waste heat can be utilized, so the energy transformation efficiency is up to 70 %. Using coal based fuel—compatible with the existing energy supply system, the IGFC system has shown potential applications in many fields, such as portable SOFC power systems, combined heat and power (CHP) systems available to buildings and large SOFC systems for distributed power generation. Given the successful application of IGFC system, the coal utilization level of China will be largely improved in the future.

5 Conclusions

Achieving a high-efficiency, clean and lower-carbon energy transition is a complicated system project, which has crucial interaction with resource, technology, environment, infrastructure, as well as the social and economic development mode. It is necessary to insist on exploring every type of desirable alternatives to find the long-term solutions and design comprehensive energy and environmental policies under current energy conditions and challenges. China needs to pay equal attention to promoting clean and high-efficiency fossil energy utilization and developing nuclear energy and renewable energy over the long run for an energy revolution. Fortunately, some positive progress in energy engineering and technologies is providing opportunities for China to change its profile of energy production and consumption, and the country needs to seize this opportunity. The road for the high-efficiency, clean and low-carbon energy transition is going to be long and challenging, but it has started.

References

1. Information Office of the State Council of China. *China's Energy Policy 2012, 2012*. http://www.gov.cn/english/official/2012-10/24/content_2250497.htm. Accessed 10 Sept 2015
2. National Bureau of Statistics of China, *China Statistical Yearbook 2015* (China Statistical Publishing House, Beijing, 2015)
3. L. Ma, J.M. Allwood, J.M. Cullen, Z. Li, The use of energy in China: tracing the flow of energy from primary source to demand drivers. Energy **40**, 174–188 (2012)
4. J.M. Cullen, J.M. Allwood, The efficient use of energy: tracing the global flow of energy from fuel to service. Energy Policy **38**, 75–81 (2010)
5. W. Ni, S. Song, M. Wang. *Developing High-Efficiency, Low-Carbon, Clean Coal in China*. http://cornerstonemag.net/developing-high-efficiency-low-carbon-clean-coal-in-china/. Accessed 10 Sept 2015

6. British Petroleum. *BP Statistical Review of World Energy June 2015*, 2015. http://www.bp.com/en/global/corporate/energy-economics/statistical-review-of-world-energy.html. Accessed 10 Sept 2015
7. Xinhua net. *China Unveils Energy Strategy, Targets for 2020, 2015*.http://news.xinhuanet.com/english/china/2014-11/19/c_133801014.htm. Accessed 10 Oct 2015
8. J. Wang, L. Feng, S. Davidsson, M. Höök, Chinese coal supply and future production outlooks. Energy **60**, 204–214 (2013)
9. Z.X. Zhang, China in the transition to a low-carbon economy. Energy Policy **38**, 6638–6653 (2010)
10. J. Hou, P. Zhang, Y. Tian, X. Yuan, Y. Yang, Developing low-carbon economy: actions, challenges and solutions for energy savings in China. Renew. Energy **36**, 3037–3042 (2011)
11. H. Bloch, S. Rafiq, R. Salim, Coal consumption, CO_2 emission and economic growth in China: empirical evidence and policy responses. Energy Econo. **34**, 518–528 (2012)
12. V.G.R.C. Govindaraju, C. Tang, The dynamic links between CO_2 emissions, economic growth and coal consumption in China and India. Appl. Energ. **104**, 310–318 (2013)
13. W.D. Ni, Z. Chen, Clean and efficient utilization of coal-key to China's low carbon economy. J. Taiyuan Univ. Technol. **41**(5), 454–458 (2010)

The Possible Rate of Transition to Lower-Carbon Housing

Philip Lloyd

Abstract Two of the challenges facing any transition to a lower carbon economy in the building sector are the questions of how rapidly the existing low-efficiency stock of domestic housing can be replaced with more efficient housing and how efficient the new housing stock can be made. This paper therefore develops a model for the replacement of the global housing stock as it ages and considers what the demand for new housing stock is likely to be. One driver will clearly be the increasing population. Another will be economic growth, which has the unexpected effect of reducing the average occupancy of homes as nations develop economically. This not only accelerates the underlying rate of increase in new housing forced by the increasing global population, but also offers opportunities for higher-value, more energy efficient homes. Moreover, economic development is usually associated with greater levels of urbanisation, which allows greater use of multi-dwelling buildings with associated improved efficiency potential. Nevertheless, the lifetime of most homes is inherently long in comparison with the apparent urgency of reducing energy demand and thus lower carbon emissions. It is concluded that, per se, more efficient housing is unlikely to play a significant part in the transition to a lower carbon world over the next 35 years until 2050.

Keywords Building sector · Housing · Energy efficiency · Demand · Population growth · Occupancy

P. Lloyd (✉)
Energy Institute, Cape Peninsula University of Technology, Cape Town, South Africa
e-mail: lloydp@cput.ac.za

P. Lloyd
SAAE, Hatfield, South Africa

K.V. Raghavan and P. Ghosh (eds.), *Energy Engineering*,
DOI 10.1007/978-981-10-3102-1_7

Table 1 Population of the world and major areas, 2015, 2030 and 2050, according to medium projection [3]

Major areas	Population (millions)			Annual growth, 2015–50 (%)
	2015	2030	2050	
World	7349	8501	9725	0.80
Africa	1186	1679	2478	2.13
Asia	4393	4923	5267	0.52
Europe	738	734	707	−0.12
Latin America and Caribbean	634	721	784	0.61
North America	358	396	433	0.54
Oceana	39	47	57	1.09

1 Introduction

The residential part of the global building sector consumed about 16 EWh of energy in 2010 [1], and that is expected to grow to about to about 30 EWh by 2050 [2] under a business-as-usual scenario.[1] This is equivalent to a growth rate of about 1.7 % per annum. Some of the growth will undoubtedly be due to population growth in Table 1.

The average home globally houses about 3.4 persons, so the growth from 7.3 billion to 9.7 billion people by 2050 implies about 700 million new homes. In addition, the existing stock of homes will age and have to be replaced. The new homes create the opportunity to introduce energy efficient structures, provided there is the political will to enforce building regulations that require energy efficiency. In addition, there are opportunities to improve the efficiency of the existing stock. Thus this paper aims to develop a model which will enable some quantification of the possibilities for moving the housing stock into a more energy efficient state, and to develop some ideas about the rate at which this could occur under different levels of policy intervention.

2 The Present Situation

At present there are approximately 1.9 billion homes in the world. This is derived from a large sample reported in Wikipedia [4], which includes the sources of the data. A portion of this sample, nations with more than 10 million households in the year reported, is given in Table 2. The population of all countries in the sample in

[1]Note that there is some discrepancy in [2]. Figure 9.3 gives what appears to be incorrect data for per capita domestic energy consumption. These data lead to an estimate of 24.3 PWh in 2010, as shown in Fig. 9.4. The data of Fig. 9.5, however, give far more reasonable data for the domestic energy consumption. These data lead to an estimate of 15.8 PWh for the global consumption..

Table 2 Households and population of nations with >10 million homes [4]

Country	Household population	Households	Household size	Year
China	1,367,820,000	455,940,000	3	2012
India	1,018,865,868	192,671,808	5.3	2001
United States	304,130,000	117,538,000	2.6	2009
Brazil	189,790,211	57,324,167	3.3	2010
Russia	142,754,098	52,711,375	2.7	2002
Japan	124,973,207	49,062,530	2.5	2005
Germany	80,645,605	40,076,000	2.0	2008
United Kingdom	64,106,779	26,473,000	2.4	2011
France	65,920,302	25,253,000	2.6	2005
Italy	60,233,948	23,848,000	2.5	2008
Vietnam	89,708,900	22,444,322	4.0	2009
Mexico	95,380,242	22,268,196	4.3	2000
Philippines	98,393,574	18,539,769	5.3	2007
Turkey	74,932,641	17,794,238	4.2	2008
Iran	77,447,168	17,352,686	4.5	2006
Ukraine	45,489,600	17,199,000	2.6	2008
Spain	46,620,045	16,741,379	2.8	2008
Korea South	45,452,526	15,887,128	2.9	2005
Ethiopia	73,302,305	15,634,304	4.7	2007
Poland	37,812,741	13,337,040	2.8	2002
Canada	35,158,304	12,437,470	2.8	2006
Argentina	39,672,520	12,171,675	3.3	2010
South Africa	53,157,490	11,205,705	4.7	2001
Colombia	41,174,853	10,570,899	3.9	2005

2011 totalled 4.8 billion, which was 68 % of the global population in 2011. The sample totalled 1.4 billion homes, or an average occupancy of 3.4 persons per dwelling.

It was observed that there was a tendency for less developed nations to have a higher occupancy than more developed nations. This was tested against all nations in the sample, using the GDP per capita as a measure of the state of development. A reasonable correlation resulted, which is shown in Fig. 1. The occupancy level in nations for which there was no data on the number of dwellings was therefore estimated from the relationship:

$$y = 35.87x^{-0.248}$$

where y is the occupancy and x the GDP per capita for purchasing power parity in 2011$. When this relationship was used to estimate the number of homes in those countries that were not included in the sample, but for which there were data on the GDP per capita, the total number of homes was estimated to be 1.86 billion with an

Fig. 1 Relation between occupancy of homes and GDP per capita

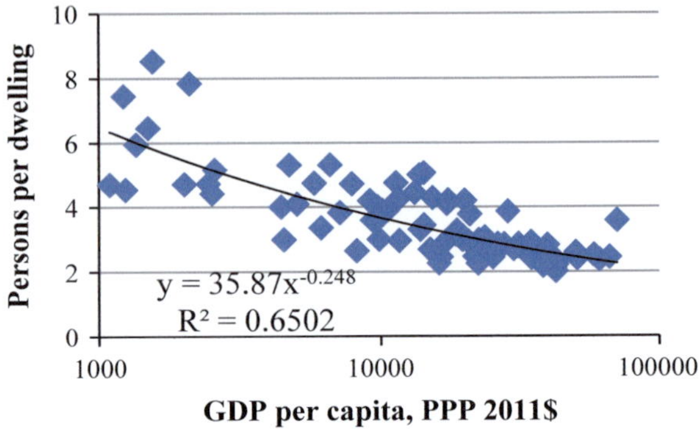

average occupancy of 3.6 persons per home. The average occupancy, 3.6, was somewhat higher than the average occupancy in the original sample, 3.4, because poorer nations were under-represented in the original sample. Additional nations with more than 10 million homes were identified, which are shown in Table 3.

Having derived a reasonably complete picture of the world, it was then possible to aggregate the various nations into regional groupings as shown in Table 4.

Table 3 Estimates of additional nations with >10 million dwellings

Country	Household population	Households	Household size
Indonesia	249,865,631	64,690,040	3.9
Pakistan	182,142,594	40,456,530	4.5
Nigeria	173,615,345	39,362,600	4.4
Bangladesh	156,594,962	30,078,625	5.2
Egypt	82,056,378	22,457,024	3.7
Vietnam	89,708,900	22,444,322	4.0
Thailand	67,010,502	19,291,457	3.5
Saudi Arabia	28,828,870	11,648,044	2.5
Algeria	39,208,194	11,231,953	3.5

Table 4 Population, households and economies of regions of the world, 2011

Region	Population[a]	Households	Occupancy	GDP/cap	GDP, 2011 $
Africa	1,064,155,026	212,431,669	5.0	4067	4.33E+12
Asia	3,927,049,240	1,036,500,083	3.8	9229	3.62E+13
Europe	823,051,563	311,607,623	2.6	27,722	2.28E+13
N America	435,107,405	152,393,278	2.9	41,578	1.81E+13
Oceana	35,357,623	11,279,277	3.1	26,957	9.53E+11
S America	471,458,411	132,223,849	3.6	12,194	5.75E+12

[a]Note that because of changes in the definition of regions, the population of the various regions is slightly different from that of Table 1

3 Future Prospects

The factors that will change the number of dwelling units over the next 35 years are:

1. An increase in the number of units required to house a growing population
2. An increase in the number of units driven by improved economies and therefore higher GDP/capita, leading to lower occupancy rates (this is also a reflection in areas such as South America and Africa of growing urbanisation and lower family sizes as a result)
3. The replacement of existing housing due to age or lack of suitability (e.g. rural homes abandoned).

The first of these is derived directly from the growth rates given in Table 1. The second is the subject of considerable guesswork. One set of pundits believes that "growth in the OECD and emerging G20 countries is likely to decelerate from 3.4 % in 1996–2010 to 2.7 % in 2010–60." [5]. Another study [6] estimates that "the estimated average real GDP growth rates for the 32 economies covered in this study over the period to 2050. Newly emerging economies such as Nigeria and Vietnam could grow at 5 % or more per annum on average over this period, whilst the growth of established emerging economies such as China may moderate to around 3–4 %. Advanced economies are projected to grow at around 1.5–2.5 % per annum in the long run, with variations reflecting different working age population growth to a significant degree." Several other studies have reached similar conclusions, so we assume the economic growth rates shown in Table 5.

The third and final factor is the rate of replacement of the existing housing stock. There is data from Europe and North America, which indicates that demolition rates are of the order of 25 % of the new construction rates, so that it was necessary to build at least 25 % more homes than were needed for new entrants into the market [7]. However, the demolition rates are precisely what they indicate—the rate at which homes are physically destroyed. Other homes are lost because they are abandoned, so that the actual loss rate is significantly higher than the demolition rate.

Table 5 Assumed economic growth rates; constant $ PPP

Region[a]	Growth rate, %/a	Region	Growth rate, %/a
CPA	3.7	PAS	2.1
EEU	1.3	POE	1.9
FSU	1.5	SAS	3
LAM	3.1	SSA	3.5
MNA	2.5	WEU	1.5
NAM	2.5		

[a]*CPA* Centrally Planned Asia; *EEU* Eastern Europe; *FSU* Former Soviet Union; *LAM* Latin America; *MNA* Middle East North Africa; *NAM* North America; *PAS* Pacific Archipelago States; *POE* Pacific OECD; *SAS* Southern Asia; *SSA* Sub-Saharan Africa; *WEU* Western Europe

Table 6 Calculation of new build for new entrants (net build) and total build allowing for demolition and abandonment (gross build)

Parameters	Region	Factor	2011	2012	2013
2.13 %	SSA	Population	843,533,804	861,501,074	879,851,047
3.50 %		GDP, 2011 $	2.48E+12	2.567E+12	2.657E+12
		GDP/cap	2.94E+03	2.98E+03	3.02E+03
		Occupancy	4.95	4.93	4.92
		Households	1.70E+08	1.75E+08	1.79E+08
		Net build		4.21E+06	4.31E+06
25 %		Gross build		5.27E+06	5.40E+06
		Demolitions		1.06E+06	1.09E+06
60 %		2011 stock	1.70E+08	1.69E+08	1.68E+08

Eventually, the lack of data was resolved by making what seemed to be reasonable assumptions about how much of the 2011 housing stock was likely still to be in use by 2050, and calculating how many extra dwelling units would have to be constructed each year, over and above the number of units needed to house the growing population. An underlying assumption is that all of the houses built between 2011 and 2050 will remain in use. The calculation for Africa for the first few years is illustrated in Table 6.

The first parameter is the annual rate of population growth, from Table 1, and the second, the assumed annual GDP growth in constant $, from Table 5. The third is the factor giving the additional number of homes to be built each year to make up for demolition or abandonment, and the fourth is the result of the calculation, the % of the 2011 housing stock still in use in 2050. First the population and the GDP are calculated each year, from which the GDP per capita is calculated, which gives the Occupancy, persons per dwelling, by the equation derived from Fig. 1. The population divided by the occupancy gives the number of households in total. The new builds (Net build) are then the difference between the number of households from one year to the next. The total of new builds (Gross build) is the Net build times (1 + parameter) where the factor is 25 % in this case. The number of demolitions and abandonments in each year is then the difference between the Gross build and the Net build, and each year the original stock is reduced by the number of demolitions and abandonments that year.

Continuing this process through 2050, the remainder of the 2011 stock is calculated, and the ratio (2011 stock remaining in 2050)/(2011 stock) then gives the fourth parameter, 60 % in this case. It is argued that a loss of 40 % of the original housing stock over 39 years is quite reasonable in the case of Africa, where the quality of the original stock is low and rapid urbanization is taking place, so there is likely to be a rapid abandonment of rural homes. The results of this exercise are summarised in Table 7.

What this indicates is that, over the next 35 years, there will be new stock needed of nearly 750 million dwellings in Asia (CPA + SAS + PAS), 350 million

Table 7 Summary of new homes required between 2015 and 2050

Region	2011 housing	2050 housing	% 2011 stock surviving	Net new dwellings occupied	Gross new dwellings built	% extra housing built each year
CPA	3.83E+08	6.34E+08	75	2.51E+08	3.47E+08	38
EEU	4.31E+07	4.07E+07	90	4.15E+06	8.36E+07	107
FSU	8.37E+07	1.13E+08	80	2.88E+07	4.56E+07	58
LAM	1.65E+08	2.64E+08	80	9.98E+07	1.33E+08	33
MNA	1.36E+08	1.98E+08	80	6.15E+07	8.88E+07	44
NAM	1.38E+08	2.05E+08	90	6.71E+07	8.09E+07	21
PAS	1.60E+08	2.68E+08	70	1.09E+08	1.56E+08	44
POE	7.26E+07	1.01E+08	90	2.88E+07	3.59E+07	25
SAS	3.20E+08	4.96E+08	75	1.76E+08	2.56E+08	45
SSA	1.70E+08	4.41E+08	60	2.71E+08	3.39E+08	25
WEU	1.82E+08	2.02E+08	90	2.09E+07	3.90E+07	87

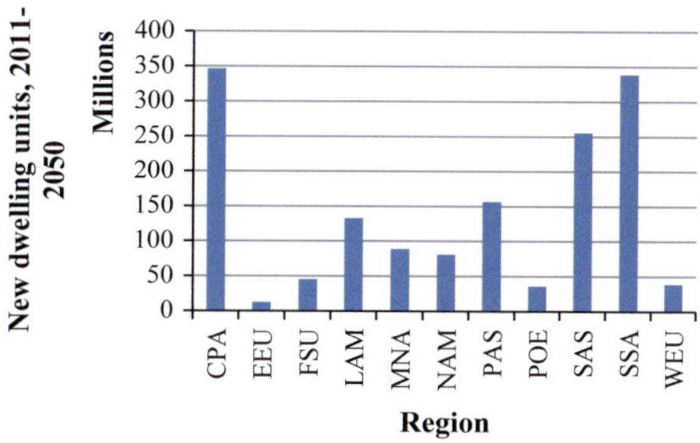

Fig. 2 Regional needs for new housing by 2050

in Africa, over 100 million in Latin America, and less than 100 million in the whole of Europe and in North America. Opportunities for introducing energy efficient housing are clearly greatest in Asia and Africa. This is illustrated in Fig. 2.

4 Business-As-Usual Energy Demands to 2050

The 2011 energy demand of households in the various regions globally is shown in Fig. 3 [2]. Over the next 39 years it may be expected to grow, particularly in less developed regions where at present access to modern energy services is limited. Implicit in this reasoning is that, in these regions, at present there is a lack of household appliances that are taken for granted in more developed regions. There will almost certainly be a rapid increase in the uptake of appliances as economic

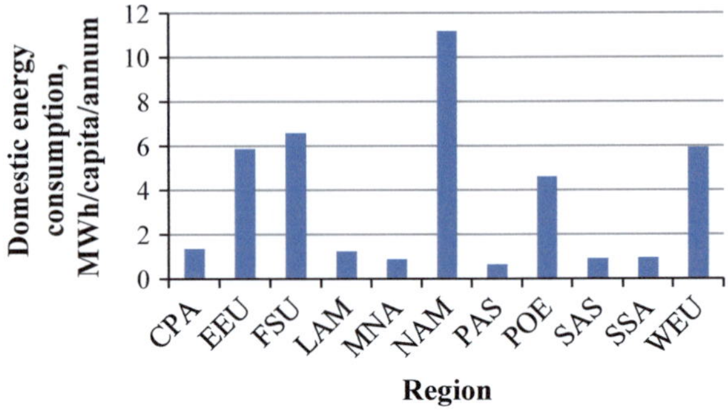

Fig. 3 Annual domestic energy consumption in global regions in 2011

development occurs, which will be a primary driver of domestic energy demand. Energy efficient housing may play a role in limiting the increase in energy consumption, which we will consider later.

There is a significant lag between the arrival of modern energy services and the full utilization of those services. When communities are first electrified, for instance, low-power demands for lighting, communication and computation dominate. This is followed by small appliances such as kettles and irons. It is typically five years before the first major appliance is acquired, usually a refrigerator [8]. After about 10 years, the recently electrified home will start to look like the average low-income home in a developed society.

The driver for development will be economic growth. There is a strong relationship between energy consumption and wealth, as indicated in Fig. 4 [9]. A very similar relationship can be traced back over the past 50 years; it is significant and persistent. High energy consuming nations tend to be those having lower ambient temperatures; conversely, lower consuming nations tend to have a tropical climate. It is possible to strengthen the relationship by correcting for the ambient temperature effect, which improves the regression coefficient R^2 to over 0.9, but this does not alter the relationship significantly.

We can therefore use this relationship to estimate the probable growth in energy demand until 2050. To allow for the slow uptake, the economic growth until 2045 will be assumed as the measure to the energy demand growth to be expected until

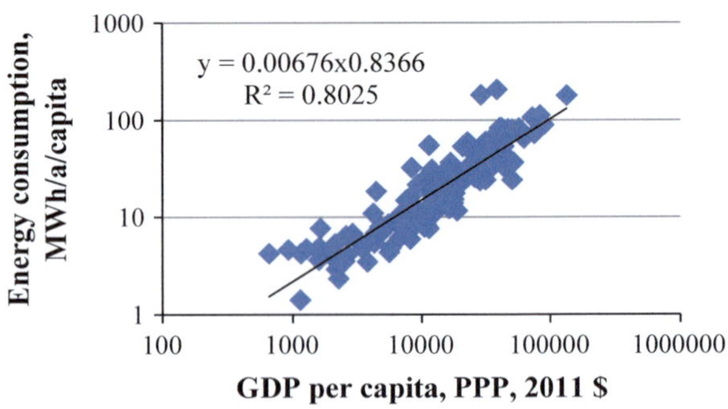

Fig. 4 Relation between GDP and energy consumption, 2011

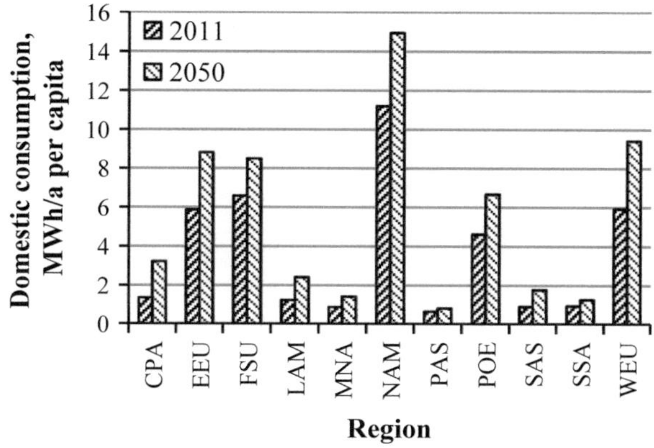

Fig. 5 Expected change in domestic energy consumption, 2011–2050

2050. Then the domestic energy consumption in each region in 2011 shown in Fig. 3 is inflated according to the relation between GDP per capita in 2011 and 2045, using the equation y = 0.00676 × 0.8366 shown in Fig. 4. The results are given in Fig. 5.

The results seem reasonable. North America will continue to dominate the per capita consumption. There will be strong growth in Eastern Europe, the Former Soviet Union and the Pacific OECD. The Centrally Planned Asian states will also see strong growth, but off a low base. Per capita consumption in the rest of the world will remain low but growing.

However, the picture changes dramatically when we look at the total energy consumed, as shown in Fig. 6. The total energy consumption in the sector will have grown from about 16 EWh in 2011 to nearly 33 EWh. The total consumption in Centrally Planned Asia will be approaching that of North America, and the Southern Asian states will be approaching that of Western Europe. The greatest proportional growth will have taken place in Sub-Saharan Africa, with over three times its 2011 consumption.

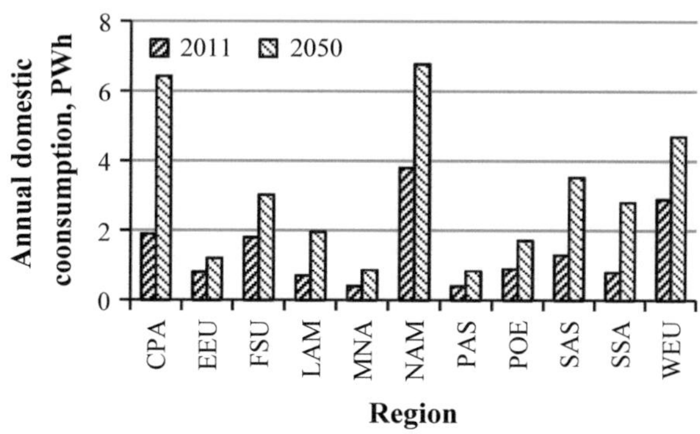

Fig. 6 Total domestic energy consumption, 2011 and 2050

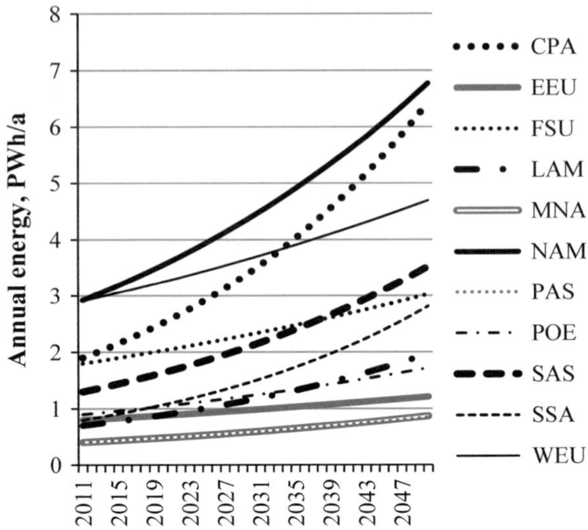

Fig. 7 Predicted growth in annual energy demand, 2011–2050

There are interesting features in the way in which the various economies grow their energy consumption in this model. Figure 7 tracks the dynamics for each region. Western Europe (WEU), the Former Soviet Union (FSU), Eastern Europe (EEU) and the Pacific Asian states (PAS) all show essentially linear growth. All have relatively low population growth and/or economic growth. The combination of high population growth and strong economic growth leads to an exponential increase in energy consumption, as so clearly shown by the Centrally Planned Asian (CPA) states. If there is to be any hope of achieving a lower carbon world, it is clearly important to find means of curbing such exponential growth.

5 A Lower Carbon World by 2050?

In previous sections, we have seen how business-as-usual models and the anticipated growth of global economies, with some large-population nations developing rapidly, are likely to lead to large increases in the demand for energy. Today there are already concerns about China and the other centrally planned Asian states not being able to foresee controlling their emissions before 2030. What we have shown is:

Those regions that already have high per capita consumption are unlikely to reduce their per capita consumptions and will probably increase their total consumption.

Those regions that are developing rapidly are likely to increase their per capita consumptions rapidly, and their population increase means that their total consumption is likely to rise rapidly.

In this light, we have to ask what assumptions might be necessary to reduce the growth in energy consumption. First, of course, is the possibility of improving the energy performance of the building envelope of new buildings. By 2050, about 1.5 billion new homes will have been built, of which 1.1 billion will be newly occupied and 0.4 billion occupied by people who previously lived in homes that were demolished or abandoned during the period. The total number of dwellings by 2050 will be about 3 billion, so that new dwellings will represent about half the stock. There are clearly opportunities for improving the energy performance of these new dwellings.

Quite wide experience has shown that it is probably possible to achieve about a 40 % reduction in energy consumption in homes without very significant changes in building practice. As a thought experiment, therefore, assume that by 2020 there is local legislation in some parts of the world requiring such a reduction in the design of every new home, and that by 2025 such legislation is widespread. Then the domestic demand pattern shown in Fig. 7 will change to that shown in Fig. 8. Comparison between the two Figures shows that there is now very low growth in seven of the regions, and that the exponential growth shown in some regions in Fig. 7 is now close to linear. The reduction in energy from 33 to 26 EWh/a represents about a 25 % reduction from the business-as-usual scenario, but it is still some 75 % above the 2011 consumption, with growth driven by NAM, CPA, WEU and SAS in particular. NAM and WEU start from a high level of consumption (and high wealth), and both enjoy quite low population growth rates, so it is evident they should strive for additional curbs on growth in energy consumption. In contrast, CPA and SAS both start from a relatively low base and have to cope with significant development issues, so it seems there is little that they could do to reduce their energy demand further.

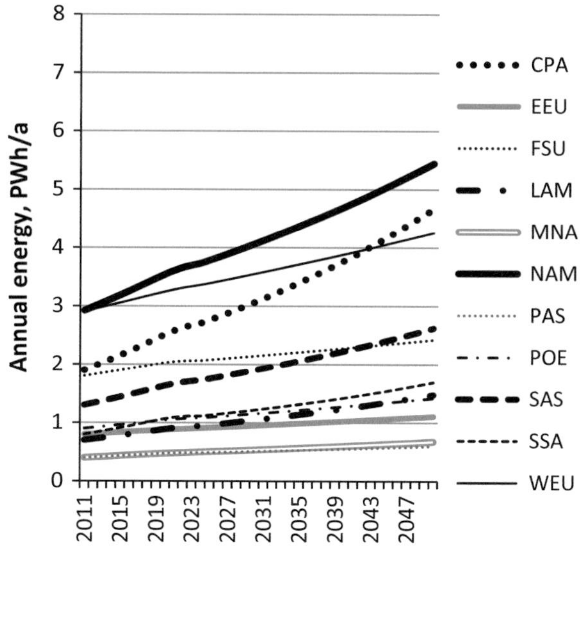

Fig. 8 Predicted growth in energy demand after implementing energy efficiency regulations for new homes from 2020

Clearly there are opportunities to enforce greater energy efficiency in existing dwellings, which could further reduce the demand, but the model foresees very aggressive action in respect of new homes, which will make up about 50 % of the total homes by 2050, so improved efficiency in existing homes will not make a great difference to the outcome.

6 Discussion and Conclusions

It seems that the possibility of achieving a lower carbon world through improvements in the energy efficiency of homes is limited. It should be possible to achieve a lower energy demand than a business-as-usual scenario, but there is still likely to be a significant growth in energy demand by 2050. The growth in the demand for homes will drive the energy growth regardless of efficiency measures.

There are two drivers for more homes, first and obviously, the growth in the population, and secondly the tendency for smaller families and thus less people per dwelling as economic development takes place. Even in developing nations, where population control measures have proved effective, there is a marked reduction in the number of people per dwelling and a rise in single-person occupancy [10].

A driver for increased energy consumption has been increasing domestic use of energy, driven by wider use of electricity in homes in developing nations. Clearly energy efficient appliances will play a part in helping to reduce the demand for energy, but even if all appliances are A-rated rather than D-rated, at best savings of the order of 50 % are achievable [11], which is insufficient to reduce 2050 consumption to less than 2011 consumption. There is already a reasonable population of A-rated appliances, so the full benefit is no longer available.

It can only be concluded that, in the area of domestic homes, the opportunities for containing the growth of energy consumption over the next 35 years are very limited; the chances of reducing the energy consumption are virtually nil. This is the only conclusion responsible engineers can reach when they look at the likely growth of housing over the next generation. There is a drive to a lower carbon world, but practical considerations indicate that, from the demand side, there is little that can be done to reduce demand significantly, particularly in the face of a strongly growing market. It is possible that low-carbon energy generation may succeed in reducing emissions on the supply side of the equation, but whether they will be able to do so at the same level of service delivery that fossil fuel-, nuclear- and hydro-powered generation currently afford is doubtful, unless cost-effective energy storage systems can be developed, something which is widely sought but tantalizingly remote at present.

Acknowledgments The Indian National Academy of Engineers is to be thanked for this kind invitation to think more deeply about the challenges presented in moving towards a lower carbon world. That the findings of this study were negative was unexpected; but it is essential that we consider negative findings if we are to avoid gross errors.

References

1. IEA Online Data Services, http://data.iea.org/ieastore/statislisting.asp. Accessed Sept 2015
2. O. Lucon, D.U. Vorsatz, in *Buildings.: Climate Change 2014: Mitigation of Climate Change*. Chapter 9 of 5th Assessment Report (ARC) of IPCC, Cambridge, UK and USA (2014)
3. United Nations Department of Economic and Social Affairs, in *Population Division*. "World Population Prospects; The 2015 Revision." New York, UN, (2015)
4. https://en.wikipedia.org/wiki/List_of_countries_by_number_of_households. Accessed Aug 2015
5. H. Braconier, G. Nicoletti, B. Westmore, Policy challenges for the next 50 years. OECD economic policy papers, no. 9, OECD Publishing, Paris. www.oecd.org/economy/Policy-challengesfor-the-next-fifty-years.pdf. Accessed Sept 2015
6. PWC, *The world in 2050. Will the Shift in Global Economic Power Continue*? www.pwc.co.uk/economics. Accessed Sept 2015
7. Carliner, M. *Replacement demand for housing* Housing economics, Dec.1990, pp. 5–9
8. P. Lloyd, in *Twenty years of knowledge about how the poor cook.* Proceedings of International Conference on Domestic Use Energy, Cape Peninsula University of Technology, 3–4 April 2012, pp. 21–30. ISBN 978-0-9814311-7-8
9. World Bank, *World Development Indicators*, Excel Spreadsheet, 2015
10. http://www.chinadaily.com.cn/china/2014-05/15/content_17508456.htm. Accessed Sept 2015
11. http://www.seai.ie/Schools/Post_Primary/Subjects/Home_Economics_JC/Appliances/. Accessed Sept 2015

Solar Cells: Materials Beyond Silicon

Soumyo Chatterjee, Uttiya Dasgupta and Amlan J. Pal

Abstract Considering India's inability to produce semiconductor grade silicon and correspondingly manufacturing of silicon solar cell modules, the prospects of other upcoming materials in third-generation solar cells have been discussed. Our outlook on solar cells based on conjugated organics, nanostructures of compound semiconductors, hybrids between two types of semiconductors, oxides, and newly-found organic-inorganic perovskites has been presented.

Keywords Materials beyond silicon for solar cells · Conjugated organics · Inorganic nanostructured materials · Organic-inorganic hybrids · Perovskites · Balance between conversion efficiency and manufacturing cost

1 Introduction

In solar cells, a potential difference is generated with a current-delivering capacity upon illumination of visible light. The sequence of steps that occur in a photovoltaic device are:

1. Generation of the charge carriers due to the absorption of photons in the active materials,
2. Separation of photo-generated charge carriers,
3. Collection of charge carriers at the respective terminals.

The prime component of a solar cell structure is therefore its absorber layer. Here the photons of incident radiation are efficiently absorbed resulting in creation of electron-hole pairs. The photo-generated electrons and holes become separated from each other followed by transport to the opposite terminals.

S. Chatterjee · U. Dasgupta · A.J. Pal (✉)
Department of Solid State Physics, Indian Association for the Cultivation of Science, Jadavpur, Kolkata 700032, India
e-mail: sspajp@iacs.res.in

K.V. Raghavan and P. Ghosh (eds.), *Energy Engineering*,
DOI 10.1007/978-981-10-3102-1_8

Most of the solar cells modules that are commercially available are based on crystalline silicon wafers. These generation solar cells have a high stability along with moderate power conversion efficiency. In a country like India, where production of semiconductor grade silicon did not somehow take place (it is said that we missed the "silicon bus"), the deployment cost per Watt (or $/W) of these solar cells would become significantly high if the all-important raw material have to be imported. Moreover, the cost-component of active material is high in silicon solar cells since the material has an indirect band gap. A thicker material is therefore required to absorb ample solar illumination.

If our inability to produce semiconductor grade silicon in India continues, we have no other option but to look for alternate materials for solar cell applications. Considering the expertise of Indian researchers in materials science, a range of conjugated organics, nanostructures of inorganic semiconductors, organic-inorganic hybrids, oxides, and perovskites, which are some of the promising and upcoming materials in this direction, could be considered since such materials can be synthesized in our country. Due to a direct band gap in most of these materials, their thin-films are sufficient to absorb solar light. Fabrication cost of solar cells based on these materials is hence less expensive. Although the efficiency of the solar cells may not match that of silicon solar modules, the lower material and manufacturing cost may make the $/W of such solar cells competitive to that of silicon solar cells.

2 Materials Beyond Silicon

2.1 Conjugated Organics

With high absorption coefficient and a low cost, conjugated organics that can be processed from solution would be a suitable replacement to silicon solar cells in fabricating organic photovoltaic devices (OPVs). Since Tang's pioneering work [1] on the bilayer heterojunction OPV reported in 1986, intensive investigation has been performed. This has led to an impressive power conversion efficiency (η) of more than 8 % [2] in such organic devices. However, due to non-geminate carrier recombination and low carrier mobility in organic semiconductors, efficiency of all organic devices is still not up to their full potential.

The concept of bulk-heterojunction (BHJ) solar cells (Fig. 1) was introduced to improve the performance of bilayer organic solar cells. In a bilayer structure, excitons generated in the vicinity of the interface (within the diffusion length of excitons) could dissociate leaving the rest to recombine in radiative/non-radiative manner. In contrast, a BHJ structure offers interfaces between electron-donor and electron-acceptor materials throughout the active layer. This leads to improved charge separation followed by carrier transport to the opposite electrodes due to bicontinuous percolative network of donor and acceptor materials in BHJs. In other words, in contrast to bilayer structures, separation of charge carriers in BHJ structures occurs all over the active layer. One of the earliest low band gap polymers used in organic BHJs is PCPDTBT, a derivative of 2,1,3-benzothiadiazole (BT). When blended with PCBM ([6,6]-phenyl-C61 butyric acid methyl ester) such devices have reported an efficiency of 3.2 % [3]. It is assumed that with a perfectly optimized PCPDTBT/PCBM active layer, power conversion efficiency in excess of 5 % could be achieved. PCBM, when mixed with another widely-used polymer P3HT, poly(3-hexylthiophene-2,5-diyl), yielded an efficiency of 6 % [4]. Clearly, there is still a scope of improvement through a control over morphology though addition of additives.

Conjugated polymers generally have a higher hole mobility as compared to that of electrons. This intrinsic imbalance in carrier mobility can be overcome by incorporation of an *n*-type inorganic semiconductor as electron-acceptor to enhance the electron transport. Such an alternative approach led to the formation of hybrid solar cells. While replacing fullerene derivatives (PCBM) as electron acceptor materials in OPVs, devices were formed with inorganic semiconductor nanocrystals (NCs), nanorods (NRs), tetrapods (TPs), and quantum dots (QDs) in a matrix of conjugated polymers resulting in solar cells based on two-types of semiconductors or hybrid materials.

Among the nanocrystals, CdS, CdSe, CdTe, ZnO, TiO_2, PbS, and PbSe were the commonly used electron acceptors. The first hybrid devices were made with CdTe nanoparticles achieving 0.05 % conversion efficiency [5]. An inappropriate band alignment was the reason behind such a poor performance. However, with further research, the efficiency of CdTe based solar cells has reached an impressive 3.2 % [6]. Use of PbS instead of CdTe extended the window of absorption up to the

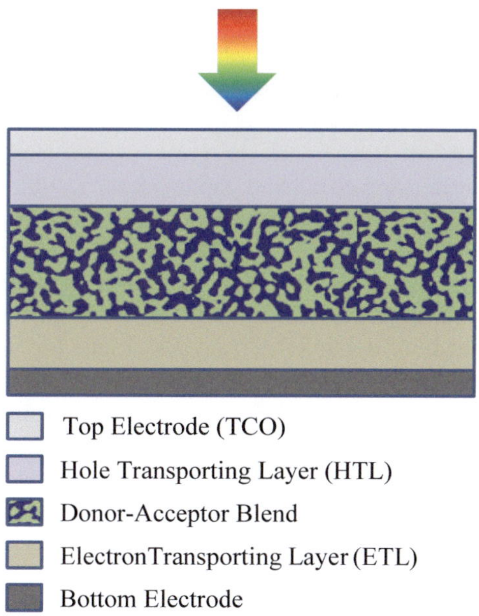

Fig. 1 Schematic diagram of device architecture of an organic BHJ solar cell

infrared region. The device efficiency with PbS has also reached 3.78 % [7]. With the ability of multiple exciton generation, the PbSe devices have performed with internal and external quantum efficiencies of more than 100 % both [8]. Other low band gap QDs such as $CuInS_2$ [9] and GaAs [10] have also shown promising output. By engineering both P3HT and QD surfaces, an efficiency of 4.1 % could be achieved with CdS nanodots [11]. Wide band gap metal oxides have also been introduced within the polymer matrix. In situ prepared ZnO QDs have demonstrated 2 % conversion efficiency [12]. With TiO_2 the efficiency has reached up to 2.2 % [13]. CdS/Cu_2S nanorods having *pn*-junction along the length have also been used charge separating channels within a P3HT matrix. Impressive conversion efficiency of 3.7 % was obtained in such hybrid BHJ devices [14]. In spite of great promises, the commercial deployment of hybrid devices did not materialize due to a low power conversion efficiency saturating at around 4 %. With appropriate QD surface and morphology engineering it is expected that the hybrid BHJs should be able to compete with existing technologies.

2.2 *Nanostructured Materials*

Over the past decade, solution-processed colloidal quantum dots (CQDs) have provided a powerful platform for the development of third generation photovoltaic cells. In addition to quantum confinement effect, ability of multiple exciton generation, size tunable optical and electrical properties etc. have made the CQDs a highly promising candidate for next-generation low-cost photovoltaic devices. Synthesis of high quality monodispersed CQDs typically involve a fast nucleation followed by a slow growth of the particles. Generally, a rapid injection of precursor solution into hot coordinating solvent triggers the nucleation [15]. With time, the molecular addition of monomeric precursor aids the growth of the nanocrystals. Interestingly, the size of such nanocrystals can be easily varied by changing the growth temperature or growth time, leading to tunable optical band gap. With decreasing size, band gap increases resulting in a blue-shift in absorption edge [16–18]. Apart from size, band gap is also dependent on constituent stoichiometries. Alloyed materials, viz. CdSeS, CdSeTe [19, 20] have shown optical characteristics that could not be achieved by their binary counterparts. Due to solution based synthesis method, introduction of small amount of impurity ions in the reaction medium provided a low-cost route for homogeneous doping. With appropriate dopant atom, electrical properties of the nanomaterials can be altered drastically. Recently *n*-type PbS QDs have been synthesized via bismuth doping resulting in an impressive photovoltaic performance [21]. Elimination of trap states at the junction by the use of same material but of alternate semiconductor-type made it possible to construct silicon-like homojunction solar cells. Schematic representation of a solar cell architecture based on nanostructured materials is shown in Fig. 2.

Another unique structure in CQDs is the core/shell configuration. Coating the QD surface with layer of another material provided a near-complete surface

Fig. 2 Schematic diagram of device architecture of a nanostructure solar cell [21]

passivation. The surface states in each QD provided the recombination pathways for CQD solar cells. This has become an effective way to reduce recombination in such core/shell structured devices. The optical properties of core/shell systems were found to depend on the core material [22]. Such systems, with a thin shell layer forming type-I band alignment, have shown high charge transfer rate [23]. A typical example of type-I core/shell system is $CuInS_2$/ZnS, which has shown nearly 7 % power conversion efficiency exceeding the performance of solar cells constructed with $CuInS_2$ without any shell material [24]. On the other hand, a type-II band alignment provides localization of different charge carriers at different region of core/shell systems. Charge separation in this type of nanometer-scale junctions reduces electron-hole recombination leading to an enhanced carrier lifetime. ZnSe/CdS [25], CdTe/CdS [26], CdS/CdSe [27], etc. are some examples of type-II core shell systems.

A significant development on material and device configuration has been made over the past decade leading to a rapid development of QD based solar cells. The technology has came a long way since the first demonstration of QD based solar cell in 2005 employing CQD and polymer layer sandwiched between ITO and Mg electrodes [28]. Later PbS or PbSe has been used as both absorbing and transporting layers in CQD based Schottky solar cells. With time, such PbS based Schottky solar cells have reached an impressive power conversion efficiency of 4 % via substitution of long chain oleic acid ligands by short chain butane di-thiols [29]. However the highest efficiencies with CQDs have been achieved via planar heterojunction structures. Materials like PbS, PbSe, and CZTS (Cu_2ZnSnS_4) have shown promising results. The highest reported QD solar cell efficiency of 10.7 % was achieved with PbS as an active material [30]. CZTS which is formed from nontoxic and earth-abundant elements has yielded a power conversion efficiency of 8.6 % [31]. BHJs between *p*- and *n*-type nanomaterials that can be considered as an inorganic analogue of organic BHJ, have shown superior conversion efficiency to the planar structure [32]. Apart from the above configurations, some unconventional routes were also considered to observe some effective results. Use of band gap gradient by employing QDs with varying size to funnel photogenerated charges to their respective electrodes turned out to be fruitful. There was a 50 % efficiency enhancement in graded structures as compared to the corresponding ungraded one [33]. It is obvious that with advancement of synthesis procedure towards defect-free QDs and fabrication of ordered compact films, the solar cells based on QDs are soon going to emerge with its full potential.

2.3 Oxides

Inclusion of oxides in a solar cell is always beneficial from the point of view of stability. This is a major advantage over the other upcoming solar materials as every solar cell would primarily operate under a harsh and ambient condition. Besides stability, most of the oxides are also nontoxic in nature. Devices based on such materials can moreover be fabricated in atmospheric condition. When oxides based on earth-abundant elements are considered in a device, the fabrication process becomes environment-friendly and cost-effective, making them ideal for large-scale photovoltaic deployment. There is however a drawback associated with the oxide materials; their band gap often does not match that of solar irradiation (Fig. 3). In general, oxide materials are classified as wide band gap semiconductors and therefore they are in general used as carrier transporting layer [34–37] in various devices. In such a group of oxide semiconductors, cuprous and cupric oxides (Cu_2O and CuO, respectively) are exceptions. These two materials have a direct band gap of 2.1 [38] and 1.4 [39] eV, respectively, which along with their natural *p*-type conductivity and high absorption coefficient make them attractive solar harvesters. Unfortunately cupric oxide has some disadvantages due to a low carrier concentration and high series resistance. These parameters limit the conversion efficiency of CuO based devices to around 1 % [39]. Cuprous oxide has excellent electrical properties, such as, long diffusion length [40] (up to several micrometers) and high mobility [41] (about 100 cm^2/Vs), leading to efficient extraction of separated carriers from the device. More recently, hole transport ability of Cu_2O has been exploited in *p-i-n* structures in the form of planar heterojunction perovskite solar cells leading to high conversion efficiency [37, 42].

Being one of the earliest photovoltaic materials, Cu_2O is being used intensively in the last decades. Failure of homojunctions [43, 44] and Schottky devices [45] as high efficiency Cu_2O based devices led to the formation of heterojunctions yielding a moderate conversion efficiency of 6.1 % [46]. The prospects with this material are

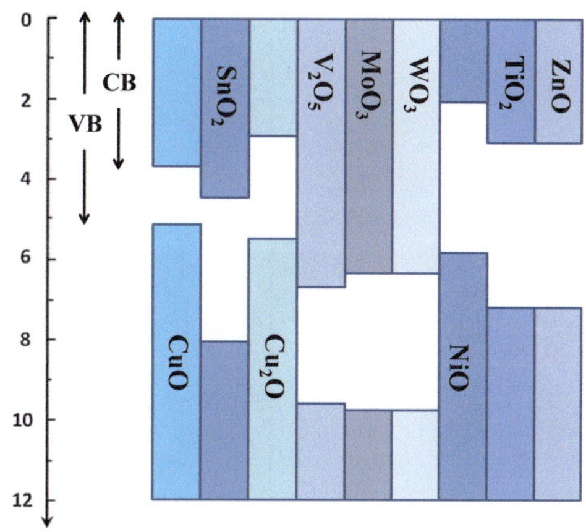

Fig. 3 Schematic energy level diagram of commonly used metal oxides

quite high if the issues like resistivity, formation of *n*-type Cu_2O, removal of oxygen deficient (*i.e.* copper rich) interfacial region, etc. can be addressed. Copper oxide based semiconductors will surely fulfill the most important criteria of the future solar cells: abundance, sustainability, non-toxicity, and ease of synthesis.

2.4 Perovskites

Use of perovskites in solar cells took a meteoric rise over a relatively short period of time as compared to most other materials in solar cells. Since the pioneering work by Kojima and his coworkers in 2009 [47], the photovoltaic conversion efficiency of perovskite solar cells has reached an almost five-fold increase with respect to the initial report. With dye-sensitized solar cells (DSSCs) device architecture in mind, the first device architecture of perovskite solar cells was very similar to that of the DSSCs. An organic-inorganic hybrid halide material, namely methylammonium lead halide ($CH_3NH_3PbX_3$; X = Cl, Br, I) has been the most popular and efficient choice as the light harvester of perovskite solar cells. Subsequent to light absorption, charge generation as well as charge extraction occurs in the perovskite layer followed by transport of holes and electrons through the hole- and electron-transport layers (HTL and ETL), respectively (Fig. 4). Due to such a two-fold role played by the perovskite in solar cells, the devices have produced promisingly high power conversion efficiency. This high output is consistent with its attractive material properties, such as high optical absorption coefficient, tunable band gap, high crystallinity, long carrier diffusion length, high mobility, ambipolar transport property, etc. [48]. The success of the organic-inorganic perovskite in carrier separation is believed to be due to the electric-dipole moment of the central moiety, namely methylammonium [49], in the compound leading to efficient charge

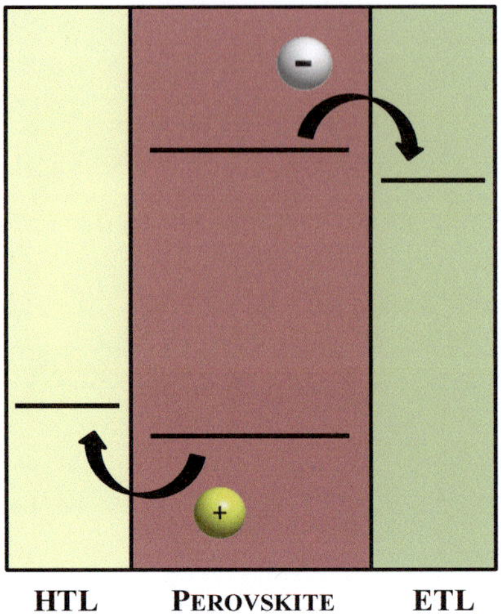

Fig. 4 Schematic diagram of charge separation mechanism in a perovskite solar cell

separation within the material itself. The origin of the dipole moment in the perovskite can be attributed to the presence of coordinate bond within the material.

Perovskite materials also have an advantage in photon energy utilization (Fig. 5), which is defined as the ratio between the open-circuit voltage (V_{OC}) and the optical band gap (E_g) during PV conversion. In standard excitonic organic solar cells, a significant amount (close to 50 %) of absorbed energy is lost during such conversion. In case of perovskite solar cells, the loss is far less. In fact, these solar cells are striving towards the same level of photon energy utilization as the current leading technologies offered by silicon or gallium arsenide.

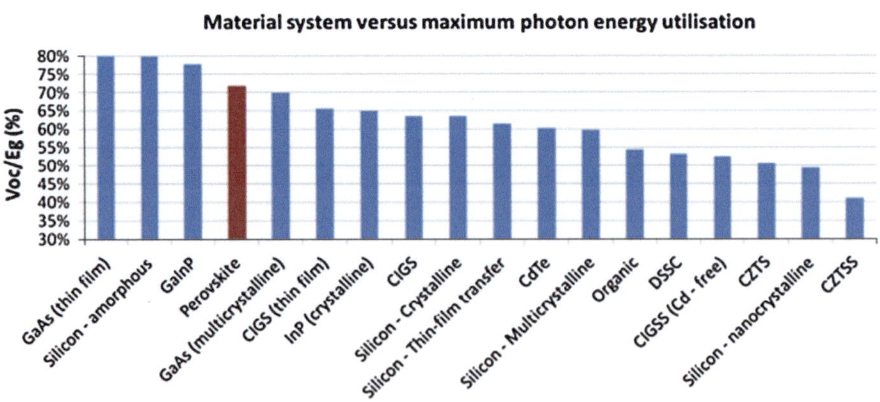

Fig. 5 The maximum photon energy utilization for common single junction solar cell materials [54]

While fabricating devices with perovskite material, an inverted structure with mesoscopic geometry was the initial choice of researchers. Using a mesoporous layer of TiO_2 as electron transporter and a layer of 2,2′,7,7′-tetrakis-(N, N-di-4-methoxyphenylamino)-9,9′-spirobifluorene (Spiro-OMeTAD) as the hole transporter, these devices has produced conversion efficiency close to 17 % [50]. Cost and complexity of perovskite solar cell fabrication could be reduced when planar heterojunction (PHJ) device architectures were put-forwarded by researchers across the globe. Based on highly uniform perovskite thin films, such inverted PHJ devices have shown promising potential reaching a conversion efficiency as high as 19.3 % [51]. However such success of inverted architectures has not deterred the quest of direct structures with perovskite materials. Recently several highly efficient perovskite solar cells have been reported having a direct structure [52, 53]. Such direct structures are not only superior to inverted ones for charge extraction but also provide the opportunity to replace Spiro-OMeTAD with a lower cost and more stable HTM, preferably an inorganic one. The direct structures (*p-i-n*) moreover allow the use of aluminum (instead of gold in *n-i-p* structures) as the top electrode thereby reducing the production cost of solar cells.

The perovskites face some sort of fabrication limitations particularly for solution processed devices. Though the best perovskite solar cells are often vacuum deposited, solution processing will ultimately lower the production cost. Replacement of costly Spiro-OMeTAD by low-cost, inorganic hole-transport materials, such as nickel oxide [55], copper iodide [56], copper thiocyanate [57] in inverted structures and formation of perovskite solar cells having a direct structure with HTMs like nickel oxide [58], copper doped nickel oxide [53], copper thiocyanate [59], cuprous oxide [37] are significant topics under investigation. These will be key routes for development of perovskite solar cells in order to achieve combination of high efficiency, long lifetime, and low manufacturing cost solar cells.

3 Conclusions

In conclusion, we have discussed the promises of several potential solar cell materials beyond silicon. The literature also presents the evolution of solar (photovoltaic) materials over the years. We have shown how newer materials and intriguing device engineering have allowed researchers to look beyond silicon as material for the future of solar cell technology. The cost efficiency and easy processability of organic-inorganic hybrid systems, multiple exciton generation ability and size tunable optical and electrical properties of inorganic quantum dots, stability and abundance of oxide materials, and high efficiency of perovskite solar cells have been discussed as potential candidates for next generation of solar cells. To reduce the cost/Watt of electricity, the importance of using earth-abundant elements as active materials for solar cells has been discussed. Also, the need for using nontoxic elements in selecting the active materials has been discussed in order to achieve environment-friendly and industrially-viable solar cells for the future.

Acknowledgements SC acknowledges DST INSPIRE Fellowship [IF140158]. UD acknowledges CSIR Junior Research Fellowship No. 09/080(0843)/2012-EMR-I (Roll No. 519699). The authors also acknowledge financial assistance from SERIIUS, DST, and DeitY projects.

References

1. C.W. Tang, 2-Layer organic photovoltaic cell. Appl. Phys. Lett. **48**, 183 (1986)
2. M.A. Green, K. Emery, Y. Hishikawa, W. Warta, Solar cell efficiency tables (Version 37). Prog. Photovoltaics **19**, 84 (2011)
3. D. Muhlbacher, M. Scharber, M. Morana, Z.G. Zhu, D. Waller, R. Gaudiana, C. Brabec, High photovoltaic performance of a low-bandgap polymer. Adv. Mater. **18**, 2884 (2006)
4. K. Kim, J. Liu, M.A.G. Namboothiry, D.L. Carroll, Roles of donor and acceptor nanodomains in 6 % Efficient thermally annealed polymer photovoltaics. Appl. Phys. Lett. **90**, 163511 (2007)
5. N.C. Greenham, X.G. Peng, A.P. Alivisatos, Charge separation and transport in conjugated-polymer/semiconductor-nanocrystal composites studied by photoluminescence quenching and photoconductivity. Phys. Rev. B **54**, 17628 (1996)
6. H.C. Chen, C.W. Lai, I.C. Wu, H.R. Pan, I.W.P. Chen, Y.K. Peng, C.L. Liu, C.H. Chen, P.T. Chou, Enhanced performance and air stability of 3.2 % hybrid solar cells: how the functional polymer and CdTe nanostructure boost the solar cell efficiency. Adv. Mater. **23**, 5451 (2011)
7. J. Seo, M.J. Cho, D. Lee, A.N. Cartwright, P.N. Prasad, Efficient heterojunction photovoltaic cell utilizing nanocomposites of lead sulfide nanocrystals and a low-bandgap polymer. Adv. Mater. **23**, 3984 (2011)
8. O.E. Semonin, J.M. Luther, S. Choi, H.Y. Chen, J.B. Gao, A.J. Nozik, M.C. Beard, Peak external photocurrent quantum efficiency exceeding 100 % via MEG in a quantum dot solar cell. Science **334**, 1530 (2011)
9. E. Arici, N.S. Sariciftci, D. Meissner, Hybrid solar cells based on nanoparticles of $CuInS_2$ in organic matrices. Adv. Funct. Mater. **13**, 165 (2003)
10. S.Q. Ren, N. Zhao, S.C. Crawford, M. Tambe, V. Bulovic, S. Gradecak, Heterojunction photovoltaics using gaas nanowires and conjugated polymers. Nano Lett. **11**, 408 (2011)
11. S.Q. Ren, L.Y. Chang, S.K. Lim, J. Zhao, M. Smith, N. Zhao, V. Bulovic, M. Bawendi, S. Gradecak, Inorganic-organic hybrid solar cell: bridging quantum dots to conjugated polymer nanowires. Nano Lett. **11**, 3998 (2011)
12. S.D. Oosterhout, M.M. Wienk, S.S. van Bavel, R. Thiedmann, L.J.A. Koster, J. Gilot, J. Loos, V. Schmidt, R.A.J. Janssen, The effect of three-dimensional morphology on the efficiency of hybrid polymer solar cells. Nat. Mater. **8**, 818 (2009)
13. Y.Y. Lin, T.H. Chu, S.S. Li, C.H. Chuang, C.H. Chang, W.F. Su, C.P. Chang, M.W. Chu, C. W. Chen, Interfacial nanostructuring on the performance of Polymer/TiO_2 nanorod bulk heterojunction solar cells. J. Am. Chem. Soc. **131**, 3644 (2009)
14. U. Dasgupta, A. Bera, A.J. Pal, *pn*-Junction nanorods in a polymer matrix: a paradigm shift from conventional hybrid bulk-heterojunction solar cells. Sol. Energy Mater. Sol. Cells **143**, 319 (2015)
15. C. Burda, X.B. Chen, R. Narayanan, M.A. El-Sayed, Chemistry and properties of nanocrystals of different shapes. Chem. Rev. **105**, 1025 (2005)
16. L. Brus, Electronic wave-functions in semiconductor clusters—experiment and theory. J. Phys. Chem. **90**, 2555 (1986)
17. I. Moreels, Y. Justo, B. De Geyter, K. Haustraete, J.C. Martins, Z. Hens, Size-tunable, bright, and stable PbS quantum dots: a surface chemistry study. ACS Nano **5**, 2004 (2011)
18. W.W. Yu, L.H. Qu, W.Z. Guo, X.G. Peng, Experimental determination of the extinction coefficient of CdTe, CdSe, and CdS nanocrystals. Chem. Mat. **15**, 2854 (2003)

19. Z.X. Pan, K. Zhao, J. Wang, H. Zhang, Y.Y. Feng, X.H. Zhong, Near infrared absorption of $CdSe_xTe_{1-x}$, alloyed quantum dot sensitized solar cells with more than 6 % efficiency and high stability. ACS Nano **7**, 5215 (2013)
20. T. Shu, Z.M. Zhou, H. Wang, G.H. Liu, P. Xiang, Y.G. Rong, H.W. Han, Y.D. Zhao, Efficient quantum dot-sensitized solar cell with tunable energy band $CdSe_xS_{(1-x)}$ quantum dots. J. Mater. Chem. **22**, 10525 (2012)
21. A. Stavrinadis, A.K. Rath, F.P.G. de Arquer, S.L. Diedenhofen, C. Magen, L. Martinez, D. So, G. Konstantatos, Heterovalent cation substitutional doping for quantum dot homojunction solar cells. Nat. Commun. **4**, 2981 (2013)
22. Y. Justo, P. Geiregat, K. Van Hoecke, F. Vanhaecke, C.D. Donega, Z. Hens, Optical properties of PbS/CdS Core/Shell quantum dots. J. Phys. Chem. C **117**, 20171 (2013)
23. H. Zhao, Z. Fan, H. Liang, G.S. Selopal, B.A. Gonfa, L. Jin, A. Soudi, D. Cui, F. Enrichi, M. M. Natile, I. Concina, D. Ma, A.O. Govorov, F. Rosei, A. Vomiero, Controlling photoinduced electron transfer from PbS@CdS Core@Shell quantum dots to metal oxide nanostructured thin films. Nanoscale **6**, 7004 (2014)
24. Z.X. Pan, I. Mora-Sero, Q. Shen, H. Zhang, Y. Li, K. Zhao, J. Wang, X.H. Zhong, J. Bisquert, High-efficiency "green" quantum dot solar cells. J. Am. Chem. Soc. **136**, 9203 (2014)
25. Z.J. Ning, H.N. Tian, C.Z. Yuan, Y. Fu, H.Y. Qin, L.C. Sun, H. Agren, Solar cells sensitized with Type-II ZnSe-CdS Core/Shell colloidal quantum dots. Chem. Commun. **47**, 1536 (2011)
26. P.T. Sheng, W.L. Li, J. Cai, X. Wang, X. Tong, Q.Y. Cai, C.A. Grimes, A novel method for the preparation of a photocorrosion stable Core/Shell CdTe/CdS Quantum Dot TiO_2 nanotube array photoelectrode demonstrating an AM 1.5 G photoconversion efficiency of 6.12 %. J. Mater. Chem. A **1**, 7806 (2013)
27. P.K. Santra, P.V. Kamat, Mn-doped quantum dot sensitized solar cells: a strategy to boost efficiency over 5 %. J. Am. Chem. Soc. **134**, 2508 (2012)
28. S.A. McDonald, G. Konstantatos, S.G. Zhang, P.W. Cyr, E.J.D. Klem, L. Levina, E.H. Sargent, Solution-processed PbS quantum dot infrared photodetectors and photovoltaics. Nat. Mater. **4**, 138 (2005)
29. K. Szendrei, W. Gomulya, M. Yarema, W. Heiss, M.A. Loi, PbS nanocrystal solar cells with high efficiency and fill factor. Appl. Phys. Lett. **97**, 203501 (2010)
30. G.H. Kim, F.P.G. de Arquer, Y.J. Yoon, X.Z. Lan, M.X. Liu, O. Voznyy, Z.Y. Yang, F. J. Fan, A.H. Ip, P. Kanjanaboos, S. Hoogland, J.Y. Kim, E.H. Sargent, High-efficiency colloidal quantum dot photovoltaics via robust self-assembled monolayers. Nano Lett. **15**, 7691 (2015)
31. W.C. Hsu, H.P. Zhou, S. Luo, T.B. Song, Y.T. Hsieh, H.S. Duan, S.L. Ye, W.B. Yang, C. J. Hsu, C.Y. Jiang, B. Bob, Y. Yang, Spatial element distribution control in a fully solution-processed nanocrystals-based 8.6 % Cu2ZnSn(S,Se)4 device. ACS Nano **8**, 9164 (2014)
32. A.K. Rath, M. Bernechea, L. Martinez, F.P.G. de Arquer, J. Osmond, G. Konstantatos, Solution-processed inorganic bulk nano-heterojunctions and their application to solar cells. Nat. Photonics **6**, 529 (2012)
33. I.J. Kramer, L. Levina, R. Debnath, D. Zhitomirsky, E.H. Sargent, Solar cells using quantum funnels. Nano Lett. **11**, 3701 (2011)
34. P. Tiwana, P. Docampo, M.B. Johnston, H.J. Snaith, L.M. Herz, Electron mobility and injection dynamics in mesoporous ZnO, SnO_2, and TiO_2 Films Used in dye-sensitized solar cells. ACS Nano **5**, 5158 (2011)
35. J.B. Gao, C.L. Perkins, J.M. Luther, M.C. Hanna, H.Y. Chen, O.E. Semonin, A.J. Nozik, R. J. Ellingson, M.C. Beard, n-type transition metal oxide as a hole extraction layer in PbS quantum dot solar cells. Nano Lett. **11**, 3263 (2011)
36. N.H. Sun, G.J. Fang, P.L. Qin, Q.A. Zheng, M.J. Wang, X. Fan, F. Cheng, J.W. Wan, X.Z. Zhao, Bulk HETEROJUNCTION SOLAR CELLS With NiO hole transporting layer based on AZO anode. Sol. Energy Mater. Sol. Cells **94**, 2328 (2010)

37. C.T. Zuo, L.M. Ding, Solution-Processed Cu_2O and CuO as hole transport materials for efficient perovskite solar cells. Small **11**, 5528 (2015)
38. K. Nakaoka, J. Ueyama, K. Ogura, Photoelectrochemical behavior of electrodeposited CuO and Cu_2O thin films on conducting substrates. J. Electrochem. Soc. **151**, C661 (2004)
39. S. Masudy-Panah, G.K. Dalapati, K. Radhakrishnan, A. Kumar, H.R. Tan, E.N. Kumar, C. Vijila, C.C. Tan, D.Z. Chi, *p*-CuO/*n*-Si heterojunction solar cells with high open circuit voltage and photocurrent through interfacial engineering. Prog. Photovoltaics **23**, 637 (2015)
40. L.C. Olsen, F.W. Addis, W. Miller, Experimental and theoretical-studies of Cu_2O solar-cells. Solar Cells **7**, 247 (1982)
41. A.O. Musa, T. Akomolafe, M.J. Carter, Production of cuprous oxide, a solar cell material, by thermal oxidation and a study of its physical and electrical properties. Sol. Energy Mater. Sol. Cells **51**, 305 (1998)
42. S. Chatterjee, A.J. Pal, Introducing Cu_2O thin films as a hole-transport layer in efficient planar perovskite solar cell structures. J. Phys. Chem. C **120**, 1428 (2016)
43. Y.K. Hsu, J.R. Wu, M.H. Chen, Y.C. Chen, Y.G. Lin, Fabrication of homojunction Cu_2O Solar Cells by electrochemical deposition. Appl. Surf. Sci. **354**, 8 (2015)
44. C.M. McShane, K.S. Choi, Junction studies on electrochemically fabricated p-n Cu_2O homojunction solar cells for efficiency enhancement. Phys. Chem. Chem. Phys. **14**, 6112 (2012)
45. L.C. Olsen, R.C. Bohara, M.W. Urie, Explanation for low-efficiency Cu_2O schottky-barrier solar-cells. Appl. Phys. Lett. **34**, 47 (1979)
46. T. Minami, Y. Nishi, T. Miyata, Heterojunction solar cell with 6 % efficiency based on an n-type Aluminum-Gallium-Oxide thin film and p-type Sodium-Doped Cu_2O SHEET. Appl. Phys. Express **8**, 022301 (2015)
47. A. Kojima, K. Teshima, Y. Shirai, T. Miyasaka, Organometal halide perovskites as visible-light sensitizers for photovoltaic cells. J. Am. Chem. Soc. **131**, 6050 (2009)
48. T.B. Song, Q. Chen, H.P. Zhou, C.Y. Jiang, H.H. Wang, Y. Yang, Y.S. Liu, J.B. You, Perovskite solar cells: film formation and properties. J. Mater. Chem. A **3**, 9032 (2015)
49. J.M. Frost, K.T. Butler, F. Brivio, C.H. Hendon, M. van Schilfgaarde, A. Walsh, Atomistic origins of high-performance in hybrid halide perovskite solar cells. Nano Lett. **14**, 2584 (2014)
50. N.J. Jeon, H.G. Lee, Y.C. Kim, J. Seo, J.H. Noh, J. Lee, S.I. Seok, o-Methoxy substituents in spiro-OMeTAD for efficient inorganic-organic hybrid perovskite solar cells. J. Am. Chem. Soc. **136**, 7837 (2014)
51. H.P. Zhou, Q. Chen, G. Li, S. Luo, T.B. Song, H.S. Duan, Z.R. Hong, J.B. You, Y.S. Liu, Y. Yang, Interface engineering of highly efficient perovskite solar cells. Science **345**, 542 (2014)
52. J. Cui, F.P. Meng, H. Zhang, K. Cao, H.L. Yuan, Y.B. Cheng, F. Huang, M.K. Wang, $CH_3NH_3PbI_3$-Based planar solar cells with magnetron-sputtered Nickel Oxide. ACS Appl. Mater. Interfaces **6**, 22862 (2014)
53. J.H. Kim, P.W. Liang, S.T. Williams, N. Cho, C.C. Chueh, M.S. Glaz, D.S. Ginger, A.K.Y. Jen, High-performance and environmentally stable planar heterojunction perovskite solar cells based on a solution-processed Copper-Doped Nickel Oxide Hole-Transporting Layer. Adv. Mater. **27**, 695 (2015)
54. S. Ossila Ltd., UK. http://www.ossila.com/pages/perovskites-and-perovskite-solar-cells-an-introduction
55. Z.H. Liu, M. Zhang, X.B. Xu, L.L. Bu, W.J. Zhang, W.H. Li, Z.X. Zhao, M.K. Wang, Y.B. Cheng, H.S. He, p-type mesoscopic NiO as an active interfacial layer for carbon counter electrode based perovskite solar cells. Dalton Trans. **44**, 3967 (2015)
56. J.A. Christians, R.C.M. Fung, P.V. Kamat, An inorganic hole conductor for organo-lead halide perovskite solar cells. Improved hole conductivity with Copper Iodide. J. Am. Chem. Soc. **136**, 758 (2014)

57. P. Qin, S. Tanaka, S. Ito, N. Tetreault, K. Manabe, H. Nishino, M.K. Nazeeruddin, M. Gratzel, Inorganic hole conductor-based lead halide perovskite solar cells with 12.4 % conversion efficiency. Nat. Commun. **5**, (2014)
58. V. Trifiletti, V. Roiati, S. Colella, R. Giannuzzi, L. De Marco, A. Rizzo, M. Manca, A. Listorti, G. Gigli, NiO/$MAPbI_{3-x}Cl_x$/PCBM: a model case for an improved understanding of inverted mesoscopic solar cells. ACS Appl. Mater. Interfaces **7**, 4283 (2015)
59. Y.X. Zhao, A.M. Nardes, K. Zhu, Solid-state mesostructured perovskite $CH_3NH_3PbI_3$ solar cells: charge transport, recombination, and diffusion length. J. Phys. Chem. Lett. **5**, 490 (2014)

Nano-Bio-Photonics: Integrating Technologies to Meet Challenges in Energy, Mobility and Healthcare

Sukhdev Roy

Abstract Taking inspiration from nature, integration of nano-, bio- and photonic technologies enables a completely new level of molding the flow of light and controlling light-matter interactions, especially to meet challenges in energy, mobility and healthcare.

Keywords Nanobiophotonics · Photoreceptors · Photodynamic therapy

1 Introduction

We face perennial challenges in not only trying to understand life, but also in preserving it and improving its quality. The three pillars of sustainability are social, environmental and economic. A system can be sustainable only when interdependences between these three ensure that it is bearable, viable and equitable, The priority areas for the industry are energy, health, mobility, IT and security, which are interrelated and offer challenges to ensure human welfare. In energy, it is to ensure an abundant, renewable, clean, efficient and cost-effective source of energy. In mobility, to provide carbon free transportation that is efficient, safe and cost-effective. In healthcare, it is to provide personalized healthcare; early diagnosis of ailments; cure for major diseases, ex. cancer, AIDS, a wide range of neurological disorders such as depression, Schizophrenia, Alzheimer's, Parkinson's, epilepsy and blindness; accessibility of services; health data security and ofcourse cost-effective treatment. Today, information technology knits through all other technologies, and the challenge is to provide ultrafast ultrahigh bandwidth information processing to meet our future requirements. However, the major challenge is to generate public opinion and encourage global co-operation and collaboration, to provide breakthroughs in science and technology, to solve these global issues.

S. Roy (✉)
Department of Physics and Computer Science, Dayalbagh Educational Institute, Agra, India
e-mail: sukhdevroy@dei.ac.in

K.V. Raghavan and P. Ghosh (eds.), *Energy Engineering*,
DOI 10.1007/978-981-10-3102-1_9

Nanophotonics, is an important interdisciplinary area that involves the study of unique light—matter interactions at the nanoscale. It offers exciting prospects for instance, size-dependent absorption and emission in quantum dots, exquisite control of excitation dynamics, field enhancement with novel optical resonances in plasmonic arrays, fabrication of exotic nanostructures such as photonic crystals to mold the flow of light, and nanotrapping with subwavelength control of field gradients. For efficient solar energy harvesting, it is necessary to ensure efficiency at all stages, i.e., absorption of light, conversion of light, collection of charged carriers and storage of energy. The best research cell efficiency at present is 46 % with concentrator based multi-junction photovoltaics. Breaking the limits of optical energy conversion requires efficient photon management [1]. Absorption of light requires harvesting solar photons distributed over a wide spectral range. This requires extending optical systems and material designs into IR, waste heat harvesting, thermophotovoltaics (TPV), reduction of thermal emission losses in concentrated solar power, spectrally selective absorbers with high absorption in the visible spectrum, low emittance in IR and narrowband thermal emitters in high-performance TPV systems. On the other hand, conversion of light requires reduction in thermalization-induced losses that involves (i) extraction of hot-charge carriers; (ii) sorting solar photons by their energy with multi-junction PV cell, PVs with up-converters, TPV system, external frequency splitters (as solar flux gets further diluted), which requires concentrators; (iii) converting photons into-higher or lower energy photons, (iv) processing them separately; and (v) removal of excess heat via non-contact radiative channels. Nanophotonics offers various potential approaches, one of them being the use of quantum dots for harvesting IR photons and carrier multiplication by UV absorption in them.

2 Nature-Inspired Nano-Bio-Photonic Applications

The aim of science is to ascertain truth by studying natural phenomena. Hence, all human activities are inspired by nature, which is true for engineering as well. For instance, plants provide inspiration for energy reservoirs, birds for airplanes, eye for a camera, brain for a supercomputer, lotus leaf for hydrophobic surfaces, termite mounds for natural coolers and dolphins for ships. Examples for the last two, are the bio-mimetic architectures of the Eastgate Mall in Harare and the Portcullis House in London, and the cleaner, quieter, more comfortable and 50 % more efficient dolphin-inspired propulsion systems. Nanostructures are also ubiquitous in nature [2]. The most common examples are butterfly wings, opals and feathers that reflect beautiful colours due to nanostructured surfaces. Nature-inspired engineering has led to the development of biomimetics, ex., fabrics designs based on morpho butterfly and adhesives based on the gecko lizard. In an interesting recent study, white butterfly has been used as a solar photovoltaic concentrator. It has been shown that at an angle of 17°, the body of the butterfly gets heated most. Using these wings on a 1 cm^2 monocrystalline Si solar cell, results in a 42.3 % increase in output power (16.8–23.9 mW) with 17-fold improvement in power to weight ratio [3]. Light is harvested in nature to provide vision by eyes, by plants to facilitate photosynthesis and by human bodies to make Vitamin D. An important area that has emerged is optofluidics [4]. It involves solar energy to facilitate photosynthesis or photocatalysis in a reactor that has nanostructures to enhance efficient fuel production in the form of hydrocarbons (oil, H_2 or isobutanol). Photobioreactors use microalgae or cyanobacteria, however the main challenge is in the uniform distribution of light throughout either in open or closed systems. Plasmonic nanoparticles have also been shown to enhance growth of photosynthetic algae. In another important study reported recently, an artificial leaf has been designed to generate H_2 directly from water and sunlight, using a two-electrode cell configuration with photoanode and cathode separated by an anion exchange membrane [5]. A prototype has been demonstrated that uses no wiring, is stable with more than 10 % efficiency under continuous operation for 40 h and generates 0.8 μl of H_2 per second.

The quality and quantity of light is different in different parts of the world. This has resulted in the evolution of different organisms that have adapted to varying conditions. Their properties have been optimized over centuries of evolution to ensure sustainability by efficiently photoconverting sunlight to sustain diverse functions. These organisms contain natural photoreceptor proteins that are abundantly found in the plant and animal kingdom. They exhibit a unique photoreaction behavior with rapid and efficient charge transfer processes on light absorption and an efficient photoresponse at low-light intensity levels. They can be isolated to form purified samples with excellent properties that can also be modified by physical, chemical and genetic engineering techniques. Moreover, they are environmentally friendly as they can be synthesized without any use of hazardous chemicals leading to environmental pollution. The major families of sensory photoreceptors found in plants are phototropins and cryptochromes (320–500 nm) and phytochromes (600–750 nm). The combinatorial interaction of receptors makes them perform various light regulated development in plants such as, flowering and circadian entrainment. One of the most sensitive photoreceptors that we need for our survival is the rhodopsin protein present in the human retina. Rhodopsins comprise two main categories, i.e, visual (photosensory pigments) and archael (light driven pumps) rhodoposins. Various kinds of archael rhodopsins have been discovered that include bacteriorhodopsin (BR) (H^+ pump), halorhodopsin (HR) (Cl^- pump), sensory, eucaryal rhodopsins and proteorhodopsin, which help organisms perform various life-sustaining functions. Micro-organisms containing rhodopsin genes range from salt flats, soil, fresh water, surface and deep sea water, glacial sea habitats to human and plant tissues as fungal pathogens. They comprise a broad phylogenetic range of microbial life that includes haloarchaea, proteobacteria, cyanobacteria, fungi, dinoflagellates and green algae. Archaea are one of the most primitive and ancient life forms on earth. BR found in the purple membrane of *halobacterium salinarum*, is one of the most outstanding photonic materials. It is a model system to understand signal transduction and undertakes the life-sustaining task of photosynthesis. On absorption of light, it undergoes conformational transitions in a reversible photocycle with intermediates spanning the entire visible spectrum. It exhibits amazing combination of properties that include high absorption cross-section and quantum efficiency, photo and thermal stability upto a temperature of 140 °C and large pH range (0–12), and flexibility in tailoring its properties through physical, chemical and biotechnological techniques. It is a smart material, as it performs multifunctions, namely, photochromic, photoelectric and phototransport. These unique properties have led to a wide range of applications that include, ATP generation in reactors, desalination of sea water, conversion of sunlight into electricity, ultrafast light and motion detection, artificial retina, real-time interferometry, pattern recognition, optical filtering, phase conjugation, optical bistability, colour classifiers, inks, photographic film, image processing, neural networks, spatial light modulators, optical displays, optical logic, slow-fast light, 2D, 3D, holographic and associative memories, harmonic generation, radiation detection, biosensors and optogenetics.

Recent nano-bio-potonic applications in photovoltaic systems have been reported with heterogeneous BR-Au nano particles (NPs) to mimic stack structure of granum, using two stacking layers with 40 nm dia AuNPs that results in a photocurrent density of 350 nA cm^{-2} with cw light [6]. Even an input flickering light modulation upto 80 Hz leads to a detectable signal. A BR-based nanostructured biophotovoltaic cell has also been demonstrated using a multilayered photoanode, with an open circuit voltage of 533 mV and photocurrent density of 1 mA cm^{-2} [7]. Many other studies have also been reported using quantum dots, maltose binding proteins and dark quenchers [8, 9]. The challenge is to enhance the efficiency through efficient charge transport from biomembrane to the metal or semiconductor surface. Hybrid BR-nano-bio systems can also be used for H_2 evolution from water. Ag/BR solid carbon cloth-supported hybrid nano-bio electrocatalyst has been recently demonstrated with low-onset overpotential of 63 mV, good durability (1000 cycles in alkaline media) and enhanced HER performance under 550 nm irradiation [10]. In an another study, TiO_2 electrodes modified with reduced graphene oxide (rGO)—BR has been demonstrated with H_2 production rates $\sim$11.24 mmol (μmol protein)$^{-1}$ h^{-1} and ninefold increase in photocurrent density [11].

All-optical switching is necessary to achieve ultrafast and ultrahigh bandwidth information processing. Ultrafast spectroscopic studies have shown that there is a blue-shifted I_{460} intermediate state that is formed in BR on absorption of a photon. Addition of Au NPs in BR leads to control of the I_{460} kinetics depending on the size and concentration of these NPs. Using a pump-probe configuration, sub-ps all-optical switching at much lower intensities and higher switching contrast, compared to Cu-Pc doped PMMA thin films ($\sim$TW/cm^2) and graphene-oxide thin films, has been shown recently [12–14]. Using multiple pump pulses, parallel logic gates have also been designed, demonstrating the application of BR based ultrafast nano-biophotonic devices and circuits for information processing. Integration of BR with ultrahigh-Q microresonators leads to enhanced functionality. In another novel study, all-optical switching of near-IR telecommunication signal, with a BR-coated silica microsphere coupled to two single mode optical fibers, to form a 2 × 2 port resonant coupler, has also been demonstrated [15, 16]. It has also been shown that cascading these switches, various kinds of low-power Boolean, reversible and reconfigurable computing circuits can be realized based on directed logic in a tree architecture [17].

Nano-bio-photonics has tremendous advantages in healthcare, in terms of targeted delivery, multimodality, multiplexity and controlled release of drugs. Hence, a wide range of applications have emerged that include, bioimaging, biosensors, laser tissue engineering and tweezers, optical activation and monitoring of therapy and optical diagnostics. These advances have led to nano-medicine, a new era in personalized medicine, with in-vivo and in-vitro diagnosis, nano-therapeutics and theranostics. Photodynamic therapy (PDT) is one of the most important treatment that uses a drug, called a photosensitizer. On exposure to a specific wavelength of light, it produces a form of oxygen that kills nearby cells. Various NPs are being studied as delivery agents, photosensitizers and energy transducers in PDT to effectively cure a wide range of diseases, especially cancer [18]. Natural photosensitizers derived from plant extracts have received much attention to avoid any side-effects [19]. Nano-bio-photonic biosensors based on optical or mechanical resonators or quantum dots and nanowires, or surface plasmons, offer unprecedented sensitivity and accuracy in detection. They offer capability of label-free single molecule/virus detection, with microfluidic integration and multiplexing, to form cost-effective lab-on-chip devices [20].

Ultra-weak biophotons emitted from living organisms that includes humans have been detected that directly depend on the stress on the metabolism of the organism that leads to the formation of reactive oxygen species [21]. It provides a means to develop a non-invasive health characterization technique. In order to develop effective healthcare, it is important to understand the basis of health and well-being. This leads to the perennial question of consciousness and the mind-body relationship. The state of mind affects the human body, whereas, the food intake or injuries to the body, have a profound effect on the mental state. It is no wonder that this century is claimed to be the century of the brain and determination of neural correlates for different functions are being intensely studied with advanced neuro-imaging techniques that includes fMRI. However, the desired spatio-temporal resolution to determine the function of a single neuron is limited. In a revolutionary development recently, natural photoreceptor proteins such as channel rhodopsin, BR, HR, LOV domains etc. have been genetically encoded in neurons to make them photosensitive and provide an exquisite control of single neurons with light. This important, challenging emerging area is called optogenetics [22]. It facilitates control of the triggering and inhibition of neuronal firing, with blue or yellow pulses of light, respectively, through an optical fiber attached to the brain of a mouse. However, such a setup limits the movement of the subject under study. To circumvent this, wireless and self-powered microelectrodes have been inserted in the brain to facilitate remote control. Another important aspect is that accurate theoretical models have been developed to support experimental studies. Although the technique has still to be applied to humans, it is a tremendous breakthrough in neuroscience with potential to not only cure major neurological disorders, but to also enhance performance and understand consciousness. Hence, using natural ultra photosensitive biomolecules along with ultrasensitive nanostructures leads to enhanced functionality for a wide range of photonic applications.

3 Conclusions

Nature functions in an integrated manner. It is therefore extremely important to integrate technologies to meet global challenges to ensure human welfare. Integration of nano, bio and information technologies, can provide sustainable solutions to energy, mobility and healthcare.

References

1. S. Boriskina, J.K. Tong, V. Ferry, J. Michel, A. Kildishev, Opt. Photon. News **48** (2015)
2. R.C. McPhedron, A.R. Parkar, Phys. Today **32** (2015)
3. K. Shanks, S. Senthilarasu, R.H.F. Constant, T.K. Mallick, Sci. Rep. **5**, 12267 (2015)
4. D. Erickson, d. Sinton, D. Psaltis, Nature Photon. **5**, 583 (2011)
5. E. Verlage et al., Energy Environ. Sci. **8**, 3166 (2015)
6. Z. Guo, D. Liang, S. Rao, Y. Xiang, Nano Energy **11**, 654 (2015)
7. R. Mohammadpour, S. Janfaza, A.C.S. Sust, Chem. Eng. **3**, 809 (2015)
8. M. Prasad, S. Roy, I.E.E.E. Trans, Nanobiosci. **11**, 410 (2012)
9. B. Mahyad, S. Janfaza, E.S. Hosseini, Adv. Colloid Interface Sci. (2015)
10. Z. Zhao et al., J. Am. Chem. Soc. **137**, 2840 (2015)
11. P. Wang, N.M. Dimitrijevic, A.Y. Chang, R.D. Schaller, Y. Liu, T. Rajh, E.A. Rozhkova, ACS Nano **8**, 7995 (2014)
12. S. Roy, C. Yadav, Laser Phys. Lett. **11**, 125901 (2014)
13. S. Roy, C. Yadav, Appl. Phys. Lett. **103**, 241113 (2013)
14. S. Roy, C. Yadav, Opt. Commun. **282**, 4435 (2011)
15. S. Roy, M. Prasad, J. Topolancik, F. Vollmer, J. Appl. Phys. **107**, 053115 (2010)
16. S. Roy, P. Sethi, J. Topolancik, F. Vollmer, Adv. Opt. Technol. **2012**, 727206 (2012)
17. S. Roy, M. Prasad, I.E.E.E. Trans, Nanobioscience **10**, 160 (2011)
18. S.S. Lucky, K.C. Soo, Y. Zhang, Chem. Rev. **115**, 1990 (2015)
19. S. Mittal, S. Roy, J.N. Srivastava, Laser Phys. **23**, 055606 (2013)
20. F. Vollmer, S. Roy, J. I.I.Sc. **92**, 233 (2012)
21. M. Cifra, J. Luminsc. **164**, 38 (2015)
22. K. Deisseroth, Nature Neurosci. **18**, 1213 (2015)
23. N. Hampp, Chem. Rev. **100**, 1755 (2000)

Carbon Dioxide to Energy: Killing Two Birds with One Stone

Samuel A. Iwarere and Deresh Ramjugernath

Abstract Carbon dioxide (CO_2) is regarded as one of the arch villains in the long debate on global warming, or what has now been more correctly termed climate change, as it accounts for approximately 85 % of the total greenhouse gases that are emitted annually. The majority of sources are from the combustion of fossil fuels. With a growing global population and consequently a growing demand for energy, there has been extensive research on alternative fuels and energy sources, as well technologies for the combustion of fossil fuels. In order to mitigate the environmental effects of carbon dioxide, numerous strategies have been proposed which focus on limitation of emissions from sources, capture, and degradation. Carbon dioxide in theory could be a potential feedstock for the production of fuel, energy, and value-added chemicals. In effect, carbon dioxide could be turned from a villain to a hero, i.e. producing energy while reducing greenhouse gases. It is therefore important that researchers continue to look for practically feasible, inexpensive, environmentally friendly, and energy efficient technologies that can utilize CO_2 by converting it into energy, liquid hydrocarbon fuels, and value-added chemicals. This review presents the current state of the art in this regard, with emphasis on technological improvements to make carbon dioxide a viable feedstock for energy and value-added chemical production.

Keywords Carbon dioxide · CO_2 utilization · Fossil fuels · Liquid hydrogen fuels · Methanol · Photoelectrochemical

S.A. Iwarere · D. Ramjugernath (✉)
Thermodynamics Research Unit, School of Engineering, Howard College Campus, University of KwaZulu-Natal, King George V Avenue, Durban 4041, South Africa
e-mail: ramjuger@ukzn.ac.za

K.V. Raghavan and P. Ghosh (eds.), *Energy Engineering*,
DOI 10.1007/978-981-10-3102-1_10

1 Introduction

Over the past few decades, there has been growing interest on the impact of carbon dioxide (CO_2) on issues related to global warming or climate change. This has led to several conferences and conventions, with the signing of pacts such as the Kyoto Protocol (United Nations Framework Convention on Climate Change) amongst the leading countries in the developed economy, with the aim of reducing he amount of CO_2 released into the atmosphere [1]. This is to mitigate increasing concentrations of greenhouse gases in the atmosphere, with approximately 85 % of the greenhouse gases emitted annually being CO_2. Combustion of fossil fuel is the major source [2] of CO_2 emitted. Therefore, different options have been researched based on the need to mitigate the environmental effects of CO_2 gas released into the atmosphere. Amongst the options considered for reducing carbon dioxide emissions is the capture of CO_2 generated from coal power plants, food processing industries, and manufacturing facilities used in the production of cements, pulp and paper, as well as in metallurgy industries [3], to mention a few. This is done to prevent CO_2 from being released into the atmosphere, but rather successfully stored in safe places, a process that is termed carbon dioxide capture and storage (CCS).

CCS is considered a technically feasible technology to reduce the amount of CO_2 released into the atmosphere. Despite the potential to significantly reduce the large amount of CO_2 emitted from thermal power plants, researchers have identified the relatively high energy penalty and capital cost as the main challenges of CCS technologies [4]. Hence, there is on-going research aimed at combining CO_2 removal with advanced coal energy conversion processes i.e. the development of an integrated gasification combining cycle (IGCC) with CO_2 capture aimed at attaining a low energy intensive capture [5]. Therefore, some researchers have classified CCS as a short/medium-term solution to CO_2 emission reduction.

It is important to mention that research is also currently being undertaken by various research groups nonincreasing the efficiency of fossil fuel-fired power plants [5] and the direct decarbonisation of fossil fuels so as to reduce CO_2 emissions from processes. However, from the viewpoint of energy production and mitigating CO_2 emissions, a research area worthy of increased efforts is the development of technologies that can re-use captured CO_2 and convert it into value-added chemicals and/or synthetic fuels. An alternative process termed carbon dioxide capture and re-use (CCR) is beginning to rapidly gain attention among researchers [6–12]. In fact, CCR is perceived to have more economic and socio-political benefits when compared with the CCS process [5]. If the captured CO_2 can be re-used as a starting material for the synthesis of valuable chemicals, energy, and important chemical intermediates instead of been considered a waste, it becomes a significant contributor towards a sustainable chemical industry. According to literature, enhanced agricultural production, enhanced oil and gas recovery, and the potential production of biodiesel using CO_2 captured from power plants are some of the opportunities for re-use of CO_2 [5]. Many more studies have been reported in literature for converting CO_2 into several industrially important chemicals using different methods and energy sources [3–12]. For example, the authors have published a review on the potential of non-thermal plasma based technologies for CO_2 dissociation as an alternative to coal gasification and natural gas reforming in the production of syngas [6]. Nonetheless, the main challenges in the re-use of captured CO_2 remains its limited influence, due to the relatively small amount of CO_2 (approximately 10 % of total annual CO_2 emitted) converted into useful chemicals and chemical intermediates by these processes in comparison with the total anthropogenic CO_2 emission of approximately 25 Gt/y [4, 5]. Therefore, more efforts in terms of research and development are needed in this regard.

This manuscript focuses on the various techniques and technologies that are being investigated for the re-use of captured CO_2 and the current economic and technical challenges associated with these technologies. Furthermore, the improvements that are needed for these CO_2 utilization techniques to become a commercially viable alternative to the burning of fossil fuels for energy and synthetic fuel production are discussed.

2 Current Technological Approach for CO_2 Utilization

The conversion of CO_2 into value-added chemicals and liquid hydrocarbon fuels have been studied extensively [3–12]. Different techniques have been employed for the utilization of waste CO_2 into industrially important chemicals such as methanol, syngas, and very importantly, liquid hydrocarbon fuels. Some of these methods have been successfully tested on a laboratory-scale and subsequently moved on to a demonstration scale, while some are still under rigorous bench-scale experimentation. A list of published articles on current state of the art methods for converting CO_2 into value-added chemicals and liquid hydrocarbon fuels is presented in Table 1.

Figure 1 shows the various pathways that can be employed in converting CO_2 into value-added chemicals and fuels. This block diagram also illustrates that the reduction of CO_2 emission can be achieved by looking at different applicable alternatives rather than one technological approach. Since the emission of CO_2 also comes from other sources such as the production of ammonia, ethane oxide, cement, iron and steel industries, aside from the power plants, it may be more practical to employ approaches that integrates different solution methods for many applications.

According to literature, CO_2 can be utilized in three major pathways [5]; two being by conversion routes and one is a non-conversion route. The two conversion routes as shown in Fig. 1, involve the conversion of CO_2 into fuel and as a feedstock for various useful chemicals. In this review, due to the limitation of space, we will categorize and present some of the technological approaches for the utilization of CO_2 as chemical, biochemical, electrochemical, photochemical, and plasma-based technologies.

Table 1 A list of some of the articles published on the conversion of CO_2 into products that act as energy storage medium

First author	Title of article	Year published
Barton [13]	Selective Solar-Driven Reduction of CO_2 to Methanol Using a Catalyzed p-GaP Based Photoelectrochemical Cell	2008
Chueh [14]	High-Flux Solar-Driven Thermochemical Dissociation of CO_2 and H_2O Using Nonstoichiometric Ceria	2010
Stechel [7]	Re-energizing CO_2 to fuels with the sun: Issues of efficiency, scale, and economics	2013
Hu [15]	Thermal, electrochemical, and photochemical conversion of CO_2 to fuels and value-added products	2013
Lebouvier [6]	Assessment of Carbon Dioxide Dissociation as a New Route for Syngas Production: A Comparative Review and Potential of Plasma-Based Technologies	2013
Meng [16]	Photothermal Conversion of CO_2 into CH_4 with H_2 over Group VIII Nanocatalysts: An Alternative Approach for Solar Fuel Production	2014
Kauffman [17]	Efficient Electrochemical CO_2 Conversion Powered by Renewable Energy	2015
Iwarere [18]	Dry reforming of methane in a tip-tip arc discharge reactor at very high pressure	2015

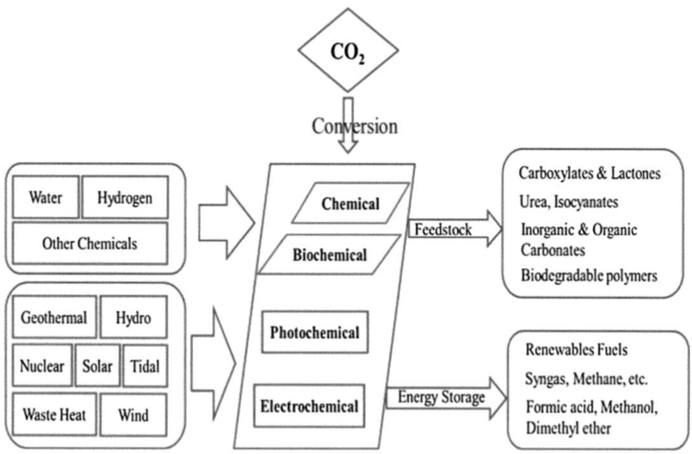

Fig. 1 Different feed stocks and energy storage pathways for the utilization of CO_2 for production of value-added chemicals and fuels

2.1 *Plasma-Based Technologies for Carbon Dioxide Utilization*

The dissociative excitation of CO_2 under thermal conditions has been studied based on the minimization of the Gibbs free energy and the result is presented in Fig. 2. As can be interpreted from the result in Fig. 2, the dissociation of CO_2 into carbon monoxide (CO) and oxygen (O_2) with no carbon deposition is thermodynamically possible at high temperature, but the process will be limited by the reaction kinetics due to the very high temperature required for the complete dissociation of the CO_2 (>5000 K). This very high temperature range is achievable in thermal plasma processes. However, extensive research has shown that the maximum energy efficiency that can be attained in processes that applies thermal dissociation of CO_2 is 43 % with the plasma acting as a high temperature heater [6]. A detailed review was presented by Rayne [19] on CO_2 dissociation using thermal and catalytic processes. Another method proposed for CO_2 dissociation with the aim of improving on the conversion and energy efficiency is the use of an electron impact process. Researchers such as Liu et al. [20] have reviewed the potential of non-thermal plasma-based technologies for the conversion of CO_2 into CO and O_2.

$$CO_2 \leftrightarrow CO + \frac{1}{2}O_2 \quad (\Delta H = 280\ \text{kJ/mol}) \tag{1}$$

CO_2 dissociation being a highly endothermic process implies that lots of energy will be required for the reactions, which will then result in low energy efficiency. However, it is reported that the energy efficiency can be improved from 43 % under thermal conditions to 61 % under nonthermal conditions [21] via the vibrational excitation of CO_2 molecules in a plasma process.

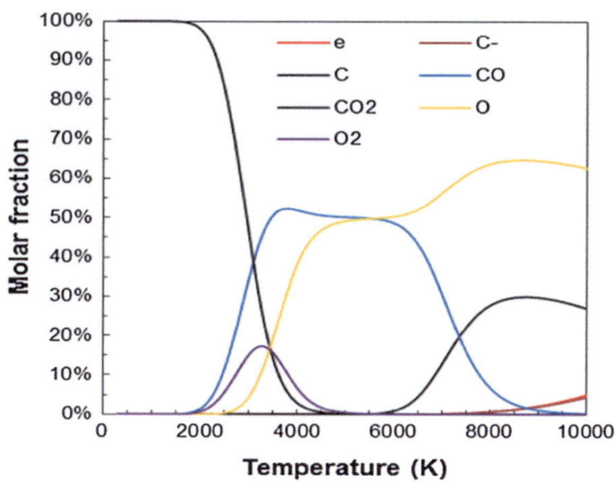

Fig. 2 Theoretical analysis of the thermal dissociation of CO_2 based on thermodynamic calculation at 1 bar [6]

$$CO_2^*(^1\Sigma^+) \rightarrow CO_2^*(^3B_2) \rightarrow CO(^1\Sigma^+) + O(^3P) \quad (2)$$

This is because vibrational excitation causes the CO_2 molecules to be stimulated from its ground electronic state to a low valence excited state according to Eq. (2) during electron impact where the CO_2 molecules dissociate into O–CO when the vibrational excitation energy is equal to the CO_2 bond energy of 5.5 eV.

The valuable O_2 and CO from the above reaction could be converted into syngas, provided the cost of producing the hydrogen (H_2) required is low enough to make the process economical. However, plasma-based CO_2 dissociation will only be competitive against conventional coal gasification and natural gas reforming for syngas production if the source of energy for CO_2 splitting is from a renewable source, e.g., solar and wind.

2.2 *Chemical and Biochemical Technologies for Carbon Dioxide Utilization*

As illustrated in Fig. 1, CO_2 as a chemical feedstock can be transformed into various spectrums of industrially useful chemicals using chemical or biochemical methods. According to statistics reported in literature, approximately 300–700 Mt/y of CO_2 emitted can be reduced by employing several chemical conversion pathways [5].

Extensive research projects are being conducted that use chemical and biochemical methods for the conversion of CO_2 to fuels [4, 5]. For instance, biomass is considered an attractive technological approach for converting CO_2 in large quantities into readily usable chemicals. As a mitigation approach to CO_2 emissions into the atmosphere as well as reducing dependence on fossil fuels, new technologies that use CO_2 to convert solar energy into biomass, which is then transformed into various renewable fuels are being promoted globally. It is expected that over a period of 30–40 years, biomass will be responsible for supplying about 5 % of the world's liquid fuel, which will result in more than 20–100 % reduction in net global CO_2 emission as compared with conventional fuels over the same period [22].

However, the high energy required for CO_2 conversion has not allowed some of the proposed conversion routes to proceed beyond laboratory-scale. Although, a small-sized reactor might be useful for pharmaceutical applications and in the synthesizing of some fine chemicals and low energy target molecules, it will not be sufficient as a storage medium for renewable energy. Therefore, continued research that aims to improve biofuel production processes is encouraged. Nonetheless, significant progress has been made in the conversion of CO_2 into valuable chemicals such as various carbonate-based materials, carbamates, salicylic acid, urea [23] and the use of biochemical methods to produce 3-hydroxypropionic [24], isobutanol, and 3-methyl-1-butanol acid [25] from CO_2.

2.3 *Electrochemical and Photochemical Technologies for Carbon Dioxide Utilization*

In the last decade, there have been significant achievements in relation to electrochemical, photochemical, and photoelectrochemical conversion of CO_2 into energy storage chemicals such as carbon monoxide (CO), dimethyl ether (DME), ethylene, formic acid, methane, and methanol [4]. A lot of emphasis is being placed on the conversion of CO_2 into energy storage chemicals because the development of an industrial-scale plants for CO_2 mitigation in this regard can significantly reduce CO_2 emission from fossil fuels while also contributing to a sustainable chemical industry. Several literature have reported improvement in selectivity and heterogeneous catalysts for CO_2 conversion into various useful products via electrochemical reduction [26]. In addition, electrochemical, photochemical, and photoelectrochemical conversion of CO_2 is preferred to other utilization pathways because renewable sources such as geothermal, solar, and wind processes can be used to drive the process, thereby making it a good electricity storage medium [5, 26].

Indeed, one of the most current research approaches is a solar-driven thermochemical process for CO_2 conversion using non-volatile mixed metal oxides. For example, Barton et al. [13] selectively converted CO_2 into methanol using a p-type gallium phosphide (GaP) semiconductor electrode and pyridinium catalyst dissolved in water with a blue light-emitting diode (LED) focused on the photoelectrochemical cell. However, their main challenge lies in the low-energy efficiency of the solar-driven reduction process. Chueh et al. [14] reported the production of fuels through a solar-driven thermochemical CO_2 and H_2O splitting cycle over cerium-oxide at high temperature. They reported solar-to-fuel efficiencies of 0.7–0.8 %, stating that the energy conversion efficiency obtained is greater than from photocatalytic approaches, with thermal losses from conductive and radiative heat transfer identified as limiting the attainment of a high efficiency.

3 Technical Challenges and Improvements of CO_2 Utilization Technologies

Several approaches have been proposed in order to reduce cost for the conversion of CO_2 into carbon-based fuels [27]. Of this, the light-driven technologies seem to be the most attractive pathways for CO_2 conversion into different high energy density components that can serve as an energy storage medium. However, to assess the potential of CO_2 utilization processes from an economically viable position, key performance indicators such as energy conversion efficiency and specific electricity consumption needs to be optimized for the plasma-based or light-driven electrochemical technology. The authors in a previous publication compared CO_2 conversion to syngas using plasma-based technologies with high CO_2 emitting technologies e.g. coal gasification and natural gas reforming based on energy cost and conversion efficiency [6].The authors are of the opinion that a plasma-based approach for syngas production from CO_2 will only be competitive with existing conventional technologies if the electrical energy source is renewable i.e. (geothermal, solar and wind) and the captured CO_2 is from a low cost power plant such as geothermal plant.

The high energy required to split CO_2 is one of the major challenges in its utilization One potential solution is to focus on reducing the energy cost in splitting CO_2 and the associated cost in converting the CO into syngas—a product that can easily be converted into industrial chemicals and liquid fuels via the F-T process. This involves investigating the cheapest alternative to producing H_2.

Let us consider the reduction of CO_2 and H_2O splitting to form high energy density products such as methanol (CH_3OH) with H_2, produced via large-scale electrolysis of water.

$$CO_2 + H_2O + \text{energy} \rightarrow CH_xO_y + zO_2 \tag{3}$$

The current state of the art approach for splitting CO_2 and H_2O is the use of a solar-driven reactor or solar-driven electrochemical cell, which makes use of specific spectrum of light from sunlight for initiating chemical reactions.

Using Fig. 3 as an illustration for energy conversion efficiency considerations; we can calculate the efficiency for methanol production for a scenario where a renewable energy source (solar) is converted to photovoltaic (PV) energy to provide the electricity for the splitting of H_2O. Assuming the maximum conversion efficiency of solar-PV is 20 %, and the PV supplies electricity for the generation of H_2 which is produced using an electrolyser with an efficiency of 65 %, then the efficiency of the PV-H_2 will be 13 %. This will lead to a total conversion efficiency of 10.4 % for CH_3OH production from PV. However, if an integrated system is considered i.e. one that has the PV cell coupled with the electrolyzer in a direct conversion approach, the total efficiency can be as high as 13.6 %. This approach of coupling a solar panel to an electrolyzer and electrochemical cell as an integrated system will help to reduce the capital cost of the electrolyzer, as well as the current density of the system and

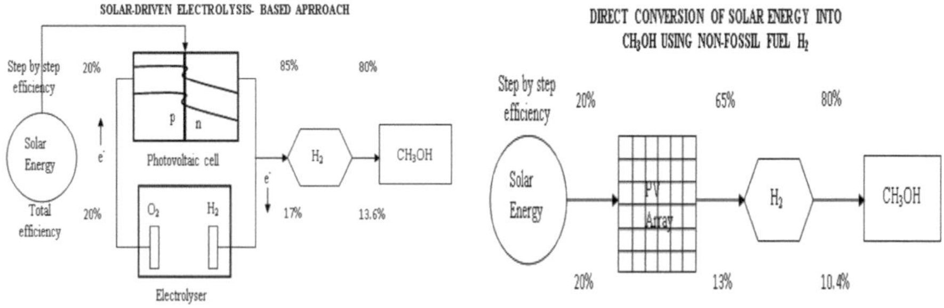

Fig. 3 Block diagrams showing photovoltaic system coupled with an electrolyzer for methanol production*(left)* and a direct photoelectrochemical system splitting CO_2 and H_2O for the production of methanol *(right)*

thus improve the overall efficiency. These scenarios provide higher efficiency than solar-to-biomass processes which are between 1 and 2 %.

From a production economics perspective for methanol, via CO_2 splitting into CO, the cost of producing hydrogen needs to be considered. Hydrogen is currently produced at 3.9 USD/kg using large scale electrolysis of water with the target cost being 2.3 USD/kg by 2020 [28].This is cheaper that 6.50–8.50 USD/kg for H_2 from nuclear and 4.5 USD/kg for H_2 from solar PV, but higher than 1.8 USD/kg of H_2 from coal.

$$3H_2 + CO_2 \rightarrow CH_3OH + H_2O \tag{4}$$

Assuming an average CH_3OH production cost of 0.9 USD/kg from coal, the cost of producing CH_3OH from PV will be higher. If the current carbon-tax price is avoided through the utilization of CO_2 in the process, the proposed amount being R120/tCO_2e in South Africa for 2016, and 168 USD/tCO_2e for Sweden in 2014 [29]. Since 1 kg H_2 will utilize 7.3 kg CO_2 according to Eq. (4), the overall H_2 cost will be reduced by approximately 1.23 USD/kg H_2 used. Based on the current value of 3.9 USD/kg for H_2 via electrolysis, the cost of producing CH_3OH from PV will be 2.67 USD/kg). If the target H_2 production cost of 2.3 USD/kg is achieved in 2020 via electrolysis of H_2O, then the real H_2cost of producing CH_3OH from PV will reduce to 1.07 USD/kg. This will make the production of CH_3OH comparable with existing processes for countries where the carbon-tax price exceed 100 USD/tCO_2e.

4 Conclusions and Perspectives

Extensive research has been carried out to demonstrate the huge potential for the utilization of CO_2 based on the volume of publications available. However, the amount of CO_2 emissions that can be reduced per year through the conversion pathways are relatively small in comparison to the 30 gigaton reported to be emitted per year (Gt/y). Several technical and political barriers such as imposing carbon tax prices are needed to make the current state of the art technologies become economically viable. In addition, transforming laboratory-scale achievements into industrial-scale need more collaboration between industry and academia. Renewable energy sources that may help to power some of the CO_2 conversion processes and reduce cost associated with high electricity consumption and low energy efficiency needs to be improved beyond its current conversion efficiency. Therefore, with the aim to effectively utilize CO_2 and reduce its emissions from burning fossil fuels, the use of multiple technologies as a well-integrated system with capability to produce several high-end products at lower energy cost compared to conventional technologies may be the future for a sustainable chemical industry.

References

1. H. Herzog, E. Drake, E. Adams, *CO_2 Capture, Reuse, and Storage Technologies for Mitigating Global Climate Change, Final Report DE-AF22-96PC01257* (MIT Energy Laboratory, Cambridge, MA, 1997)
2. U.S. EPA. Inventory of U.S. Greenhouse Gas Emissions and Sinks 1990–2007; United States Environmental Protection Agency: Washington, DC, 2009
3. S. Abanades, A. Le Gal, CO_2 splitting by thermo-chemical looping based on $Zr_xCe_{1x}O_2$ oxygen carriers for synthetic fuel generation. Fuel **102**, 180–186 (2012)
4. I. Ganesh, Conversion of carbon dioxide into methanol—a potential liquid fuel: Fundamental challenges and opportunities (a review). Renew. Sustain. Energy Rev. **31**, 221–257 (2014)
5. DNV, Carbon Dioxide Utilization: Electrochemical Conversion of CO_2—Opportunities and Challenges, Research and Innovation, Position Paper 07, 2011
6. A. Lebouvier, S.A. Iwarere, P. d'Argenlieu, D. Ramjugernath, L. Fulcheri, Assessment of carbon dioxide dissociation as a new route for syngas production: A comparative review and potential of plasma-based technologies. Energy Fuels **27**, 2712–2722 (2013)
7. E.B. Stechel, J.E. Miller, Re-energizing CO_2 to fuels with the sun: issues of efficiency, scale, and economics. J. CO_2 Utilization **1**, 28–36 (2013)
8. M. Aresta, A. Dibenedetto, A. Angelini, The changing paradigm in CO_2 utilization. J. CO_2 Utilization, **3–4**, 65–73 (2013)
9. J. Kim, C.A. Henao, T.A. Johnson, D.E. Dedrick, J.E. Miller, E.B. Stechel, C.T. Maravelias, Methanol production from CO_2 using solar-thermal energy: process development and techno-economic analysis. Energy Environ. Sci. **4**, 3122–3132 (2011)
10. J. Kim, T.A. Johnson, J.E. Miller, E.B. Stechel, C.T. Maravelias, Fuel production from CO_2 using solar-thermal energy: system level analysis. Energy Environ. Sci. **5**, 8417–8429 (2012)
11. G. Ghadimkhani, N.R. de Tacconi, W. Chanmanee, C. Janakyab, K. Rajeshwar, Efficient solar photoelectrosynthesis of methanol from carbon dioxide using hybrid CuO–Cu_2O semiconductor nanorod arrays. Chem. Commun. **49**, 1297–1299 (2013)
12. M.-A. Courtemanche, M.-A. Legare, L. Maron, F.-G. Fontaine, A highly Active Phosphine-Borane Organocatalyst for the reduction of CO_2 to methanol using hydrocarbones. J. Am. Soc. **135**, 9326–9329 (2013)
13. E.E. Barton, D.M. Rampulla, A.B. Bocarsly, Selective solar-driven reduction of CO_2 to methanol using a catalyzed p-GaP based photoelectrochemical cell. J. Am. Chem. Soc. **130**, 6342–6344 (2008)
14. W.C. Chueh, C. Falter, M. Abbott, D. Scipio, P. Furler, S.M. Haile, A. Steinfeld, High-flux solar-driven thermochemical dissociation of CO_2 and H_2O using Nonstoichiometric Ceria. Science **330**, 1797–1801 (2010)
15. B. Hu, C. Guild, S.L. Suib, Thermal, electrochemical, and photochemical conversion of CO_2 to fuels and value-added products. J. CO_2 Utilization, **1**, 18–27 (2013)
16. X. Meng, T. Wang, L. Liu, S. Ouyang, P. Li, H. Hu, T. Kako, H. Iwai, A. Tanaka, J. Ye, Photothermal conversion of CO_2 into CH_4 with H_2 over group VIII nanocatalysts: an alternative approach for solar fuel production. Angew. Chem. **126**, 11662–11666 (2014)
17. D.R. Kauffman, J. Thakkar, R. Siva, C. Matranga, P.R. Ohodnicki, C. Zeng, R. Jin, Efficient electrochemical CO_2 conversion powered by renewable energy. ACS Appl. Mater. Interfaces **7**, 15626–15632 (2015)
18. S.A. Iwarere, V.-J. Rohani, D. Ramjugernath, L. Fulcheri, Dry reforming of methane in a tip-tip arc discharge reactor at very high pressure. Int. J. Hydrogen Energy **40**, 3388–3401 (2015)
19. S. Rayne, Thermal Carbon Dioxide Splitting: a summary of the peer-reviewed scientific literature. Available Nat. Prec. (2008). doi:10.1038/npre.2008.1741.2
20. C.-J. Liu, G.-H. Xu, T. Wang, Non-thermal plasma approaches in CO utilization. Fuel Process. Technol. **58**, 119–134 (1999)

21. A.V. Eletskii, B.M. Smirnov, Dissociation of molecules in plasma and gas: the energy. Pure Appl. Chem. **57**, 1235–1244 (1985)
22. International Energy Outlook 2010, U.S. Energy Information Administration, DOE/EIA-0484, 2010
23. D. Chaturvedi, S. Ray, Versatile use of carbon dioxide in the synthesis of carbamates. Monatsh. Chem. **137**, 127–145 (2006)
24. E.I. Lan, D.S. Chuang, C.R. Shen, A.M. Lee, S.Y. Ro, J.C. Liao, Metabolic engineering of cyanobacteria for photosynthetic 3-hydroxypropionic acid production from CO_2 using Synechococcus elongatus PCC 7942. Metab. Eng. **31**, 163–170 (2015)
25. H. Li, P.H. Opgenorth, D.G. Wernick, S. Rogers, T.-Y. Wu, W. Higashide, P. Malati, Y.-X. Huo, K.M. Cho, J.C. Liao, Integrated Electromicrobial conversion of CO2 to higher alcohols. Science **335**, 1596 (2012)
26. E.B. Cole, A.B. Bocarsly, Photochemical, Electrochemical, and Photoelectrochemical Reduction of Carbon Dioxide. M. Aresta, ed. by *Carbon Dioxide as Chemical Feedstock* (Wiley-VCH Verlag GmBH& Co. KGaA, 2010), pp. 291–316
27. C. Agrafiotis, M. Roeb, C. Sattler, A review on solar thermal syngas production via redox pair-based water/carbon dioxide splitting thermochemical cycles. Renew. Sustain. Energy Rev. **42**, 254–285 (2015)
28. K. Ayers, Economical production of hydrogen through development of novel, High-Efficiency Electrocatalysts for alkaline membrane electrolysis, department of energy hydrogen and fuel cells program, Fiscal Year 2014 Annual Progress Report, 2014
29. Putting a Price on Carbon with a Tax—World Bank. Accessed from www.worldbank.com/content/

Multi Scale Optimization of Building Energy

Stephane Ploix

Abstract This paper is an introduction to smart-energy buildings. It states the user-centric energy management problem and points out how optimization can yield useful feedbacks to occupants. It focuses on the modeling issues and on the different kinds of optimizations useful for providing daily support to occupants. An application example is presented.

Keywords Energy management · Building · Optimization · Model transformation

1 Introduction

Nowadays, almost everyone knows that energy has to be saved. Scientific researches focus on the design of renewable energy production means and on improving efficiency of energy processes but usage is disregarded. Energy sobriety has a very significant impact, for instance, in 2012, in India, a citizen consumed 760 kWh whereas in US, it was 12,954 kWh. Building sector has a very strong impact on energy consumption for two reasons: it accounts for about 30 % of the CO_2 emissions in the world but also, in countries like France, 65 % of the electricity consumption is dedicated to buildings i.e. building sector is a key actor for grid demand response policies. In Europe, new standards appeared to lay down better efficiency and renewables in buildings. Nevertheless, for the period 1976–2012 in France, consumption for HVAC has decreased by 30 % whereas the consumption of the so-called other usage of electricity has increased by 90 %. It arises that end-users do not feel concerned by everyday savings because of unconscious daily routines and missing information. This paper is an introduction to smart-energy

S. Ploix (✉)
G-SCOP Lab CNRS UMR 5272, Grenoble Institute of Technology,
46 Avenue Felix Viallet, 38000 Grenoble, France
e-mail: stephane.ploix@grenoble-inp.fr

K.V. Raghavan and P. Ghosh (eds.), *Energy Engineering*,
DOI 10.1007/978-981-10-3102-1_11

buildings. It states the user-centric energy management problem and points out how optimization can yield useful feedbacks to occupants. It focuses on the modeling issues and on the different kinds of optimizations useful for providing daily support to occupants. An application example is presented.

2 Context

Before going deeper into building energy management and optimization, let's characterized the overall energy management problem. Because of constraining building standards, buildings and their appliances are becoming more and more efficient. As a result, consumption related to human activity is relatively much bigger than before. In addition, demand response in both electric and heat grids [1] lead to variable tariffs that the occupants of living areas have to take into account in their everyday life: their actions and activities now do matter and it is becoming much more complex that before. Existing building automations compute relevant set-points for HVAC systems according to occupant calendars but it does not help for most decision occupants have to take. Indeed, the following questions may arise. When and how long opening windows, shutters, flaps or internal doors, taking account weather and energy cost? Where and when to go to a specific room? When to switch the HVAC system? What should be the set points? When to switch complementary cooling or heating appliances? When to use an electric appliance taking into account variable energy tariffs? In modern buildings, when to store/restore energy? Nowadays, complexity in buildings, where people spend most of their life, appeals decision-aided systems interacting with building managers and occupants to support their actions sometimes with automations and sometimes with suggestions of actions. These systems are called energy managers. Because of thermal inertia and of daily life cycle, energy managers have to be able to anticipate the upcoming day, at least; not being just reactive at current situation.

Living area systems are quite unique because they are highly human-machine cooperative systems. Indeed, the physical and control part provides services to occupants but occupants are also part of the system and influence it a lot with their own behavior. Human activities cannot be neglected and are part of the system state. It is composed of:

- **context** related to time, weather conditions, energy costs, heat gains per zone but also occupant current positions and activities
- **controls** related to doors, windows, flaps and shutters positions, configurations of the HVAC system and other electric appliances
- **reactions** related to indoor temperature and air quality, and to satisfaction of occupants regarding services provided by electric appliances.

Future building energy management systems may cover a large set of applications [2, 3]. It can support retrospective analyses of past periods by estimating and correlating non-measured actions and variables such as occupancy, usage of windows or heat flows through windows for instance. Using simulation means, it can extrapolate future states depending on hypothetical controls or replay past situations changing the controls. To develop all these applications, energy management systems have to embed knowledge and data models of the living area system; they have to be equipped with learning, estimation, simulation and optimization capabilities.

Next sections illustrate some of these functionalities that rely on optimization processes.

3 Optimization to Solve Energy Management Problem

Finding the best controls to satisfy an overall objective is an inverse problem that can be solved using optimization processes. The first step is to state the problem to solve.

In [4], we have proposed to model occupant comfort expectations by satisfaction functions σ_i where 100 % means fully satisfied regarding service *i*, and 0 % means unacceptable. Figure 1a represents a satisfaction function modeling thermal comfort preferences. Figure 1b represents 2 different ways of modeling the user preferences regarding the ending time of a temporary service such as a washing appliance.

In the same paper, it has been shown that the weighted sum of all satisfactions can represent an overall satisfaction: weights determine the relative importance of each.

Maximizing overall satisfaction is not enough because energy cost has also to be taken into account. Problem is then a multi-objective problem. We have proposed to transform the problem into a mono-objective problem using weights corresponding to occupant profiles in order to reduce the computations. Indeed, let's consider a simple living area model with 20 variables, including 10 binary variables modeling

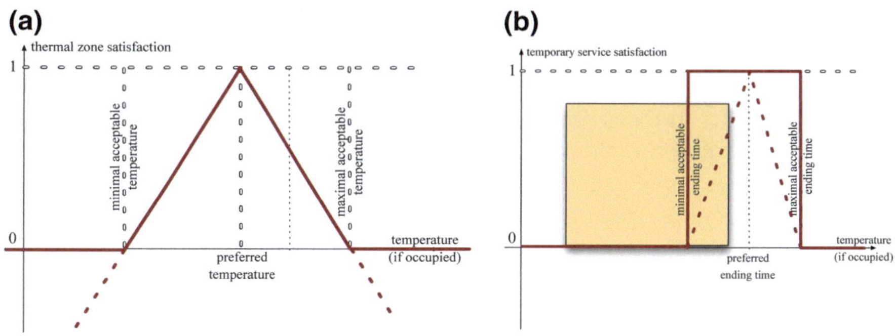

Fig. 1 Two typical satisfaction functions

different configurations of the area. Solving the anticipative problem for the next 24 h with a 1 h time resolution results in 480 variables. Most optimization algorithms will fail facing such a problem. We have proposed to automatically linearize the optimization problems. It results in an increase of the number of variables but powerful Mixed Integer Linear Programming can be used. It solves in 20 s a problem related to a building platform named PREDIS/MHI, located at Grenoble Institute of Technology, with 1500 variables for each daily anticipation. It has also be applied to a complex building prototype named CANOPEA with more that 20,000 variables within 3–4 min. Multi-agent approaches for solving singular problems, i.e. sub-problems that cannot be linearized for instance, have been presented in [5, 6].

Optimization not only appears in the computation of optimal controls satisfying objectives: it should also be used to adjust the parameters of the system model.

4 Optimization to Learn Building Behavior

Two main streams are followed to learn building behavioral models. The first one relies on universal models such as ARX, ARMAX polynomial models or neural networks. Patterns of said universal models are chosen so that the parameter values can be easily determined using for instance linear regressions. A drawback of using these models is that their structures are usually far from physical knowledge. In building systems, there are strong nonlinearities regarding for instance impact of airflows on thermic (ventilation). In addition, building systems usually also contain discrete variables that cannot be represented by a universal structure. Nevertheless, models coming from physics have also a major drawback: there are singular and nonlinear. Consequently, all the parameters cannot be estimated. A common symptom of this limitation is the sliding of the solution to non-meaningful value area for the estimated parameters. Another common symptom is the non-ergodicity of the results: a slightly different initialization can lead to drastically different parameter estimations. These conclusions have been shown in [7]. It also points out that living area models depend on phenomena usually not measured such door and windows, number of occupants and their activities that impact metabolisms. A solution has been proposed: weighting error with an inverse function of the error: it amounts to give more importance to sample where model fits well the measurements in order to focus on period with less uncertainties.

In order to solve the sliding and non-ergodicity issues, [8] has proposed a meta-optimization approach. It consists in driving a process of sequential optimizations used for parameter estimations. Let's call Θ_i the current parameter values; each parameter can vary within an acceptable interval. The first step is to select the parameters to be optimized by scanning each parameter direction, searching whether a minimum exists thanks to several simulations. Figure 2 shows when a parameter is considered as adjustable. Then, only the adjustable parameters are modified by the optimization process thanks to an nonlinear optimization where

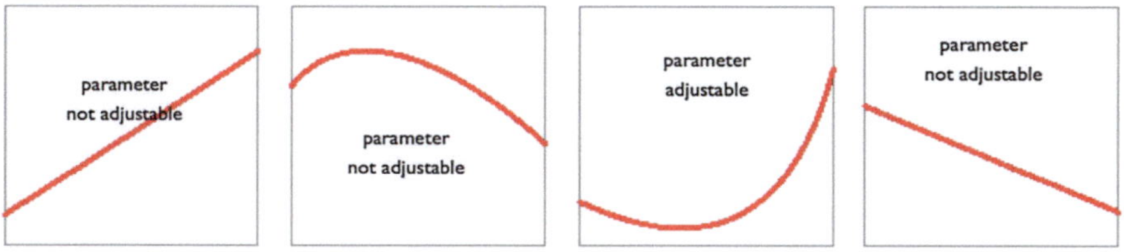

Fig. 2 Estimation error along a given parameter direction in the parameter search space

adjustable parameter values are constrained with corresponding value domains. If a parameter value comes to be stuck to a bound of its value domain, it is reinitialized to its initial value. If the parameter values have already been found, a set of parameter values is randomly drawn within the acceptable value domains to diversify and continue the process. Then, Θ_{i+1} is obtained and the process continues. This algorithm is the only one that proved to be reliable for building systems where data are usually poor because most of the dynamics correspond to a 24 h period. Frequency corresponding to 12 h also appears in spectrum analysis but less.

5 Other Considerations Related to Building Energy Management

The paper focuses on 2 kinds of optimizations but, as mentioned at the beginning of the paper, building energy management requires much more applications such as estimations and simulations. All these applications rely on the same descriptive model of a building but each one uses it in a different way.

Mixed Integer Linear Programming for optimizing anticipative energy management requires a linear in the state variables, and a causal model with an objective function related to comfort. Estimation of model parameters relies on a causal and nonlinear in the parameters formulation of the model where inputs are state variables and output are parameter values. Simulation and estimation use context and control variables of the state to determine reactions of the living area provided the parameter values are known. All these applications use a common descriptive model but in different ways. Transformation from descriptive model to applicative models can handle manually but it is not realistic to do it for each living area. Quasi-automatic transformations have been proposed in [9]. It relies on a Computer Algebraic System with transformation recipes based on primitives like simplify or solve. The causal orientation is performed thanks to the Dulmage-Mendelsohn algorithm.

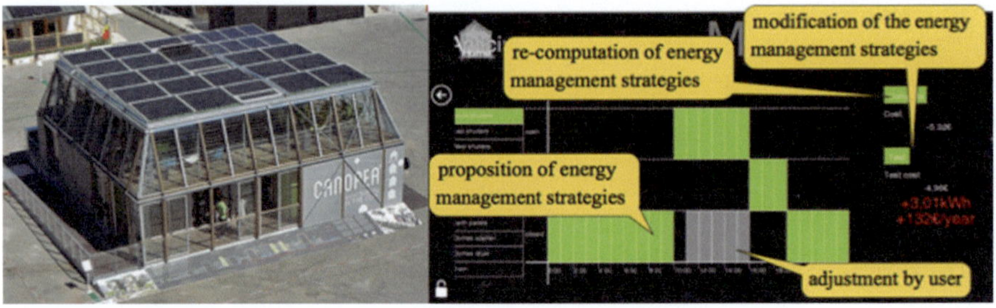

Fig. 3 CANOPEA and the user interface of the energy management system

6 Conclusions and Future Works

Previous works have been implemented in an energy-smart building prototype named CANOPEA that won the Solar Decathlon Europe 2012 worldwide contest [10]. The energy management system is based on a virtual representation of the building system including envelope, domestic appliances and technical appliances. MILP optimization has been used to generate anticipative management strategies. Model can also be used for simulation thanks to model transformation processes. Figure 3 shows how the interactions with occupants have been managed but a lot more remains to be done. Indeed, considering all the services in a living area, automation is quite limited because of the number of required sensors and actuators. It is also limited for the sake of acceptability. Involving people is a probably better option. It has lots of advantages: people are mobile sensors and actuators. They better know their expectations. Future generation of smart energy management will certainly interact more with people and will play an accompanying role to support people in reaching their own targets.

References

1. H. Farhangi, The path of the smart grid. IEEE Power Energ. Mag. **8**(1), 18–28 (2010)
2. K. Basu, L. Hawarah, N. Arghira, H. Joumaa, S. Ploix, A prediction system for home appliance usage. Energy Build. **67**, 668–679 (2013)
3. A. Kashif, S. Ploix, J. Dugdale, X. Binh Le, Simulating the dynamics of occupant behaviour for power management in residential buildings. Energy Build. **56**, 85–93 (2013)
4. L.D. Ha, H. Joumaa, S. Ploix, M. Jacomino, An optimal approach for electrical management problem in dwelings. Energy Build. **45**, 1–14 (2012)
5. H. Joumaa, S. Ploix, S. Abras, G. De Oliveira, A MAS integrated into home automation system, for the resolution of power management problem in smart homes, in Elsevier, editor, *The 1st Conference and Exhibition Impact of Integrated Clean Energy on the Future of the Mediterranean Environment*, 2011
6. H. Joumaa, S. Ploix, M. Jacomino, A mixed (centralized/distributed) solving approach for energy management problem in dwelling, in *The 2012 International Conference on Artificial Intelligence*, Las Vegas, USA, 2012

7. S. Sarabi, S. Ploix, H. Minh Le, H.-A. Dang, F. Wurtz, Assessing the relevance of reduced order models for building envelop, in *Building Simulation* (BS 2013), 2013
8. A. Le Mounier, B. Delinchant, S. Ploix, Determination of relevant model structures for self-learning energy management system, in *Building Simulation and Optimization 2014*, London, U.K., 23–24 June 2014
9. Q.D. Ngo, Y. Hadjsaid, S. Ploix, B. Parisse, U. Maulik, Toward the automation of model transformation for optimized building energy management, in *2013 IEEE Conference on Clean Energy and Technology (CEAT)*, pp. 336–341. IEEE, 2013
10. Y. Hadjsaid, B. Lechat, C. Latremoliere, S. Ploix, Generating global energy management problems: application to the canopea prototype, in *Building Simulation (BS2013), IBPSA World Conference*, Chamb´ery, France, 25–28 August 2013

Emerging Construction Materials for Energy Infrastructure

V.K.R. Kodur, M.Z. Naser and P.P. Bhatt

Abstract This paper presents the role of emerging construction materials in the development of energy infrastructure. Recent advancements in construction materials is reviewed. Specific examples of application of high performance concretes, high performance steels, and wood composites in enhancing the efficiency of infra-structure needed for generating energy from conventional fuel sources is illustrated. In addition, application of specialized materials such as fiber reinforced composites, shape memory alloys and phase-change materials in energy installations for harnessing power from non-conventional energy sources is discussed.

Keywords Energy · Emerging materials · Construction materials · Sustainability · Energy infrastructure

1 Introduction

Since ancient times, advances in the development of materials and energy production have defined human social, technological and political aspirations. The discovery of electricity and related technologies in the nineteenth and twentieth centuries started the industrial revolution and continues to shape the path of societal advancement of near future [1]. With large increase in world population, combined with improved living standards of people in developing countries, the need for energy has significantly increased. According to the International Energy Agency (IEA), based in Paris, the world's energy demand will increase from 12 billion tonne oil equivalents (t.o.e.) in 2009 to about 17 billion t.o.e. by 2035 [2]. For generating this levels of energy efficient construction and maintenance of energy infrastructure is required.

V.K.R. Kodur (✉) · M.Z. Naser · P.P. Bhatt
Department of Civil and Environmental Engineering, Michigan State University, East Lansing, MI, USA
e-mail: kodur@egr.msu.edu

K.V. Raghavan and P. Ghosh (eds.), *Energy Engineering*,
DOI 10.1007/978-981-10-3102-1_12

Fossil fuels with more than two thirds of the current and projected energy generation, continue to be the main source of energy. Even a developed country like United States is heavily reliant on fossil fuels that are non-renewable energy sources coming from crude oil, natural gas and coal. In addition to fossil fuels, energy in recent years is being generated from renewable resources such as solar, wind, hydro and geothermal sources. Renewable energy is generally defined as energy that comes from resources which are naturally replenished on a human timescale [3]. Energy generated from solar, wind, hydro and geothermal sources can often be produced at lower costs and can lead to harnessing and production of almost free-toxic/waste energy. However, these nonrenewable energy resources are not being fully utilized due to lack of adequate infrastructure required to harness energy.

In order to ensure efficient harvesting and generation of energy from different resources, state-of-the-art infrastructure built with advanced materials is needed. This paper presents the role of emerging construction materials in the development of energy infrastructure. Recent advancements in construction materials is discussed and specific examples of application of high performance concretes, high strength steels, and wood composites in enhancing efficiency of infrastructure needed for harvesting energy from conventional fuel sources is provided. Further, the application of specialized materials such as fiber reinforced composites, shape memory alloys and phase-change materials in energy installations for harnessing power from non-conventional energy sources is highlighted.

2 Infrastructures for Energy

Built infrastructure is needed for efficient generation, storing and transmission of power from conventional or renewable sources of energy. The infrastructure requirements for harnessing renewable and non-renewable sources of energy can be grouped under three broad categories, infrastructure for extraction of raw materials/fuels, conversion and storage of energy, and transmission of generated power to the end user. In this section, a brief overview of energy related infrastructure needed for mining/processing of raw fuels or materials, generating and storing power, as well as transmission to the end user, is discussed.

2.1 Fossil Fuels—Coal, Crude Oil and Natural Gas

Fossil fuels form the main source of energy over the past century and continue to serve as the main fuel to generate electricity in thermal power plants and also as basic fuel in transportation and industrial sectors. The infrastructure needed for the extraction and processing of coal, crude oil and natural gas include, oil platforms/mines with mining/drilling facilities (to mine coal, oil and gas), refinery units for distilling the oil into various fractions, pipelines for supplying oil and natural gas to end user or thermal power plants. The infrastructure requirements for power generation from coal and crude oil are thermal power stations and transmission lines to distribute generated power to end users.

2.2 Hydroelectric Energy

Hydroelectricity is generated utilizing kinetic energy of flowing water. Hence, to provide a continuous and reliable source of water supply, dams or reservoirs to store water, water or hydro turbines, generators, and long distance power lines to transmit power is needed. Hybrid systems are envisioned for use of wind, solar, or tidal/wave power to pump water behind dams and then release the water to generate electricity when it is needed.

2.3 Nuclear Energy

Nuclear energy is the energy stored at atomic level and is extracted via controlled nuclear reaction such as nuclear fission of radioactive elements in a nuclear reactor. The nuclear power plant is similar to that of a conventional thermal power plant, comprising of steam turbines, generators, transmission lines and storage units, however, uses a nuclear reactor to provide heat and generate steam. A nuclear reactor primary consists of number of energy installations including fuel assemblies, moderator, coolant, control rods, and supporting structure for fuel rods, all housed in horizontal or vertical vessels.

2.4 Solar Energy

Solar energy refers to utilizing radiant light and heat from the sun to harness energy using a wide range of advanced technologies such as solar heating, photo-voltaics, solar thermal energy, solar architecture and artificial photosynthesis [4]. One of the main advantages in harnessing solar energy is the minimal need for large infrastructure facilities, such as complex power plants. The required infra-structure to generate power (electricity) from solar energy comprise of limited to freestanding towers and solar cell panels. In such facilities, the sunlight can be concentrated to small beams using different installations i.e., parabolic trough, linear Fresnel collector, solar dish and solar power tower.

2.5 Wind Energy

Wind power is extracted using wind turbines (or mills), which convert kinetic energy of wind to electric power [5]. The needed infrastructure to harness wind energy include wind turbines, turbine tower, turbine blades, foundation for resting the turbine tower, sensors to monitor and rotate the turbine blades towards wind direction, and cables to transmit electricity to transmission lines. In order to generate feasible amounts of energy, large number of turbines are usually used in multiple grids (arrays). Such cluster of turbines is referred to as wind farms. It should be noted that wind farms can be constructed on both on-shore and off-shore locations.

2.6 Geothermal Energy

Geothermal energy is the thermal energy generated and stored in Earth since temperatures from the surface to the center of the Earth warm up by 17 to 30 °C for every kilometer in depth, reaching 5100 °C in the inner core (nearly the temperature of the sun). The infrastructure required for generating geothermal power includes injection wells, boreholes, heat exchangers, steam turbines, pipes for carrying fluid (water or butane), generator, condenser, cooling tower, transformer and transmission lines.

2.7 Transmission Lines

The electricity (power) generated in power plants is to be supplied to the end user after conversion at a substation using transmission lines. At the power plant, energy is converted into a set of three alternating electric currents, called three-phase power. After conversion of power to higher voltage (to minimize electricity lost during transmission), electricity is delivered to consumers. The main energy installations needed for transmission of generated power in power plants to end users include, transmission towers (poles), special insulators, conductors, vibration damper to reduce stresses on vibrating conductor to avoid structural damage due to vibration on transmission lines, as well electric substations, and transformers.

3 Materials for Energy Infrastructure

Conventional construction materials used in civil infrastructure is also used in the construction of much of energy infrastructure discussed in previous section. In the last three decades, continuous research in the area of construction materials has led to significant improvements in the properties of conventional materials, as well as in the development of new materials. These advances have led to development of high performance concretes, high strength steels, and wood composites mainly to overcome some of the drawbacks associated with their parent materials. In addition, new specialized materials such as fiber reinforced polymer composites (FRP), shape memory alloys (SMA) and phase change materials (PCM), with targeted properties, have been developed for use in specialized applications. This section provides a brief review of the advancements in the conventional construction materials and development of specialized materials. Special application of these high performing materials in energy infrastructure is illustrated.

3.1 Conventional Construction Materials

The conventional materials used in energy infrastructure mainly include concrete, steel and timber. In recent years, these construction materials have been undergoing continuous improvements in their properties and use of these improved types in energy installations can enhance efficiency in energy infrastructure and also reduce overall carbon foot print.

3.1.1 High Performance Concrete

Concrete is one of the most widely used materials in civil infrastructure due to its low cost, locally available ingredients (materials) and technologies for making concrete. Over the last three decades, there have been significant research and development in improving the properties of concrete. This has led to production of new types of concrete, with better properties, often referred to as high strength concrete (HSC), fiber reinforced concrete (FRC), self-consolidated concrete (SCC), high performance concrete (HPC) and ultra-high performance concrete (UHPC).

Most of the improved versions of concrete are typically characterized by higher strength, enhanced workability, improved durability, high resistance to chemical attacks, and thus are better suited for energy installations. For instance, fiber reinforced concretes, with lower permeability, higher tensile strength, better thermal resistance and superior energy absorption properties can overcome some of the problems related to seepage in dams and reservoirs needed for generating hydro-electric power. Reactive powder concrete provides a biological shield (an effective barrier) against the lethal radiation in nuclear reactors and has been successfully utilized in Hinkley Point nuclear power station in Britain [6]. High strength concretes with superior strength and durability properties has been successfully used in offshore oil platforms to overcome durability problems that is common with the use of conventional concrete (example: Hibernia oil platform in Canada). Moreover, steel fiber reinforced concrete can be highly effective in nuclear power plants (i.e., containment building "reactor chamber" and be in specialized foundations) to achieve better thermal resistance and resistance to radiation leakage.

3.1.2 High Performance Steel

Steel is widely used in construction and industrial sectors due to numerous advantages properties such as ease of fabrication, ductility, reusability, reliability, durability and efficiency that offers over other materials. Steel used in construction applications is classified into different categories based on chemical composition, mechanical properties, and fabrication process. These categories include carbon steel, high strength steel (HSS), high-strength low alloy steels (HSLA), heat-treated carbon steels (HTC), and heat-treated constructional alloy (HTCA) steels. In recent years, a number of alloy steel wherein variety of elements such as carbon, copper and magnesium, in small amounts, are added with ferrous to improve specific properties such as strength, corrosion resistance, thermal resistance and weldability.

The use of alloy steels in energy installations; such as oil and gas wells and platforms, wind turbines (blades and towers), and utility poles for transmission lines can help overcome many of the drawbacks of conventional steel. Thermo-mechanically high strength (TMHS) steel can be used in offshore mining infra-structure due to its high toughness (at low temperature) and excellent weld-ability [7]. This type of steel has been utilized in the design of oil platform Ekofisk II in the off-shore of Norway [7]. Processed high strength steel is also being used in other energy installations such as, pipelines, storage tanks, wind turbine components and electromagnetic shields.

3.1.3 Wood Composites

Timber is widely used in residential and commercial low rise buildings, as well as in some energy infrastructure such as transmission poles. Conventional wood poles is highly susceptible to environmental deterioration i.e., moisture, shrinkage and swelling. In recent years some of these drawbacks with conventional wood are overcome through wood-plastic composites (WPCs), which are made by combining wood fibers and thermoplastics resins [8]. WPCs are sustainable and eco-friendly as they are made of recycled plastics and waste products generated in wood industry. These WPCs possess properties similar to polymer composites and are highly resistant to rot, decay, and marine borer attack.

The improved versions of wood composites can be highly attractive in energy installations, such as poles carrying transmission lines, and towers for wind turbines and solar panels. Recently, wind turbine towers as high as 140 m was constructed by a German based firm (TimberTower). This firm plans to extend the height of the tower to 200 m in the near future.

3.2 *Specialized Materials*

Continuous research and development in the area of materials over the last few decades has led to the development of new and specialized materials such as fiber polymer composites (FRP), shape memory alloys (SMA) and phase change materials (PCM). These materials originally developed for specific applications in the industrial and civil infrastructure can be highly attractive in energy related installations.

3.2.1 Fiber Polymer Composites

Fiber reinforced polymers (FRP), originally developed in 1960s for aerospace applications, has emerged as a viable material in repair and retrofitting of civil infrastructure in the last 20–30 years [9]. FRP composites, made of different fiber (carbon/glass/aramid) and matrix/resin (unsaturated polyester, epoxy, vinylester, phenolic, polyurethane) combinations have unique properties i.e., high strength and high stiffness, light weight, ease of installation, high corrosion resistance and improved durability.

FRP composites can be used in a number of infrastructure installations in all sectors of energy. For instance, in case of oil and natural gas exploration, FRP pipelines can be an attractive alternative to steel pipelines in demanding environments like off-shore pipelines to transport oil and high volume/high pressure natural gas [9]. This is due to the fact that FRP pipelines, unlike steel, are highly resistant to corrosion, durable, light in weight and feasible in long sections with ease of handling and connectability. FRP composites are currently replacing steel in wind turbines and turbine blades. FRP's are also considered as a viable alternative to timber, steel and pre-stressed concrete for transmission towers and wind mills as they are non-conductive, non-corrosive, lightweight, easy to install and possess high ductility characteristics.

3.2.2 Shape Memory Alloys

Shape-memory alloy (SMAs), developed in the last two decades, are alloys that are smart, adaptive and when deformed returns to its pre-deformed shape upon heating. These SMA's can be readily deformed by applying an external force, and will contract or recover to its original form when heated beyond a certain temperature either by external or internal heating; or other relevant stimuli such as a magnetic field.

SMA's can be highly effective in energy installations, such as couplers for pipelines carrying oil and natural gas, as well as in heat engines. Recently, SMA's are also used as bracing wires for frames and connectors in off-shore oil platforms. Owing to their high power density, high actuation force and/or their strain capability, SMA wires can be embedded in wind turbine blades to provide load control surface [10]. In addition, SMA's are extensively used as ground isolators, energy dissipaters, dampers, and as structural reinforcement in energy installations such as dams and nuclear plants, especially those designed and constructed in seismic/tsunami active areas.

3.2.3 Phase-Change Materials

A phase-change material (PCM) is a substance with a high heat of fusion which upon melting and solidifying at a certain temperature, is capable of storing and releasing large amounts of energy [11]. Heat is absorbed or released when the material changes from solid state to liquid state and vice versa; thus, PCMs are classified as latent heat storage (LHS) units. These materials have been used in conventional structural systems (especially in specific bridges) to reduce the number of freeze/thaw cycles experienced by a bridge deck. By incorporating PCMs, pore solution in cementitious mixtures (materials) can be better controlled which can minimize temperature-induced cracking of concrete.

PCM's are used in energy infrastructure applications for regulating temperature and storing of heat. For instance, cubicle houses located in Lleida (Spain) were constructed with concrete mixed with PCMs and showed improved heat storage capacity and stabilized system temperature [12]. PCM's can also be incorporated in concrete mixes used in construction of energy installations such as nuclear and geothermal power plants. Construction of energy infrastructure projects require large amount of concrete casting (mass concrete), thus addition of PCM to such mixes can prevent high hydration (curing temperature) peaks, which can produce higher compressive strength and durable concrete [11]. Similar concrete mixes (with embedded PCMs) are used in foundations for wind turbines and solar panels placed in regions where there is significant changes in daily temperature. Finally, PCM materials can be added to the back of solar panels to help maintain the operating temperature within 40–80 °C.

4 Research Needs

Although there have been significant advances in material technology, still further research is needed to enhance properties and characteristics of conventional and emerging materials to better suit energy installations. Some of the research needs for various materials are briefly discussed here.

Further research is needed to improve fire resistance properties of newer types of concrete which exhibit poor performance under fire conditions and are also susceptible to fire induce spalling. Similarly, supplementary studies are needed to develop solutions for overcoming temperature induced degradation of strength, stiffness and creep properties of high performance steel once exposed to fire. Use of wood composites in power transmission poles can be further enhanced once optimum solutions relating to quality control, fiber alignment, and fatigue resistance properties is developed.

In order to efficiently employ FRP composites in energy infrastructure applications, brittleness of FRP's is to be overcome. The wider use of SMA for wind turbine blades is currently hindered due to its poor fatigue properties over time. To efficiently use phase-change materials in energy installations, further investigation especially in areas such as design, fabrication, production and synergy with other types of materials is warranted. The use of SMAs and PCMs (especially those derived from silicon, telluride and selenide/sulfide products) in the design of small and large scale of photovoltaic (PV) solar panels can be enhanced if effective implementation techniques are developed.

5 Summary

Energy is of vital importance for social, economic and overall well-being of society. In order to fulfill growing energy needs, built infrastructure made of advanced and emerging materials is a key requirement for generation, storing and transmitting of harnessed energy from conventional and renewable fuel resources. The use of advanced construction materials can provide a sustainable and efficient path to energy infrastructure for harvesting, conversion and storing of energy. However, there are few limitations associated with some of these materials which hinders their full scale use in energy installations. Thus, additional research is warranted to better enhance material properties and performance especially when subjected to harsh environmental exposure and loading conditions such as those encountered in energy sector.

References

1. S. Chu, A. Majumdar, Opportunities and challenges for a sustainable energy future. Nature 488 (2012)
2. National Academy of Engineering and National Research Council, *America's Energy Future: Technology and Transformation* (National Academies Press, 2009)
3. M. Jacobson, M. Delucchi, *A Path to Sustainable Energy by 2030* (Scientific American, 2009)
4. B. Parida, S. Iniyan, R. Goic, A review of solar photovoltaic technologies. Renew. Sustain. Energy Rev. **15** (2011)
5. R. Lacal-Arantegui, *2013 JRC Wind Status Report* (Publications Office of the European Union, Luxembourg, 2014)
6. D.J. Naus, *The Effect of Elevated Temperature on Concrete Materials and Structures—A Literature Review* (Oak Ridge National Laboratory, Oak Ridge, TNl, 2008)
7. F. Schroter, Steels for modern steel construction and offshore applications. 10th Nordic Steel Construction Conference, Copenhagen (2004)

8. J. Kim, K. Pal, *Recent Advances in the Processing of Wood-Plastic Composites* (Springer, Berlin, 2010)
9. M. Ehsani, C. Peña, *Rehabilitation of FRP Pipelines with Minimum Downtime* (Water World, 2009)
10. A. Lara-Quintanilla, A. Hulskamp, H. Bersee, A high-rate shape memory alloy actuator for aerodynamic load control on wind turbines. J. Intell. Mater. Syst. Struct. (2013)
11. M. Hunger, A.G. Entrop, I. Mandilaras, H. Brouwers, M. Founti, The behavior of self-compacting concrete containing micro-encapsulated phase change materials. Cement Concrete Composites **31**, (2009)
12. C. Castellon, M. Nogues, J. Roca, M. Medrano, L.F. Cabeza, *Microen-capsulated Phase Change Materials (PCM) for Building Applications* (Proc. Ecostock Conf., USA, 2006)

Energy Efficient Power Generation and Water Management Through Membrane Technology

S. Sridhar

Abstract Membrane technology has made rapid strides as a vital research area in chemical engineering for having exhibited tremendous potential in solving challenging separation problems such as industrial wastewater management for solvent recovery and water reclamation along with energy generation through fuel cell technology besides low cost compact plant designs for drinking water purification.

Keywords Membrane technology · Energy conservation · Fuel cell · Wastewater treatment · Drinking water purification

1 Introduction

Escalation in human population has resulted in a paucity of conventional non-renewable fuels, as well as depletion in ground and surface water resources. Therefore a worldwide pursuit for utilization of alternative energy sources which are sustainable and eco-friendly has become a source of concern [1]. Membrane technology has the potential to contribute immensely to both energy and water sectors. Membrane based fuel cell has emerged as a budding clean energy source with high energy efficiency, less consumption of resources, compactness, light weighted, reduced noise and zero emission of pollutant [2]. In the water sector, membranes have found increasing application for reclamation of water and valuable organics from industrial effluents. Reverse osmosis, ultrafiltration and more recently nanofiltration are being widely used as less energy intensive processes for producing clean and safe drinking water from sources contaminated by excess fluoride, arsenic, nitrate, iron, turbidity and pathogens.

Generally Nafion (perfluorinated membrane) manufactured by Dupont is used as a commercial membrane for fuel cells but its high cost of manufacture, expensive

S. Sridhar (✉)
Membrane Separations Group, Chemical Engineering Division,
Indian Institute of Chemical Technology, Hyderabad 500007, India
e-mail: sridhar11in@yahoo.com

K.V. Raghavan and P. Ghosh (eds.), *Energy Engineering*,
DOI 10.1007/978-981-10-3102-1_13

platinum based catalyst, low water retention and poor thermal and mechanical strength of the membrane limits its usage at higher operating temperatures. Consequently, significant research has been done to amend the Nafion membrane or develop new PEM membrane with desired properties for fuel cell applications over the years. Polymers with aromatic backbone such as polyethersulfone (PES), polyetheretherketone (PEEK), polybenzimidazole (PBI), etc. and their subsequent modified forms prepared by sulfonation or blending have also been studied [3–5]. In the present work an indigenously prepared polyelectrolyte complex blend membrane is reported (Figs. 1 and 2).

The integration of membrane processes with conventional methods such as evaporation or distillation could result in huge energy savings. The oil drilling

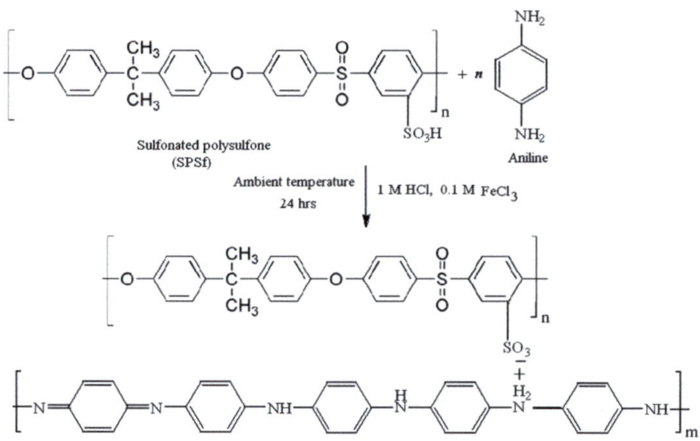

Fig. 1 Schematic representation of the polymerization of aniline on the surface and within the pores of sulfonated polyethersulfone membrane

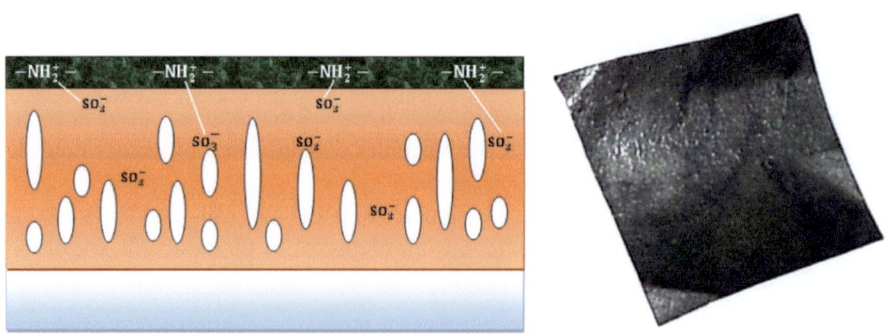

Fig. 2 Physical structure and photograph of thin film composite PANi-SPES membrane for PEMFC and DMFC applications

effluent from a coastal region was subjected to a combination of sand filtration and reverse osmosis to reduce the load on evaporator. In another case study electrodialysis-distillation hybrid process was applied to recycle DMSO solvent from different bulk drug effluents which contained hazardous or corrosive salts along with water [6]. Textile industries that produce acrylic fiber ingredient for making polyester use aqueous sodium thiocyanate as the solvent for spinning the fiber. Nanofiltration has enabled decolorization of the used solvent at half the energy consumed by conventional gel filtration process [7]. For purification of drinking water, the designing of compact systems for RO, NF and UF has resulted in low capital investment whereas the synthesis of hollow fiber membranes has enormously brought down the operating costs due to high surface area per unit volume ratio and low operating pressures in systems such as hand operated ones for flood affected regions [8].

1.1 Experimental

1.1.1 Fuel Cells

Polyethersulfone polymer was treated with sulfuric acid for the sulfonation. The sulfonated polyethersulfone (SPES) of 15 % (wt/v) were dissolved in *N*-Methyl-2-pyrrolidone solvent to get homogeneous bubble free solution. The prepared solution cast on non-woven fabric support to make SPES thin film composite membrane by phase inversion technique. Aniline was sprayed or spread on the SPES porous membranes then they were immersed into a bath which contained a $FeCl_3$ (0.1 M) solution in water and 1 M HCl. The polymerization of aniline at the surface and within the pores can be easily observed due to a change in color of membranes from white to black transiting light and dark green.

1.1.2 Wastewater Treatment

Figure 3 represents the schematic of a highly compact skid mounted Nanofiltration (NF) system which costs only Rs. 3.25 Lakhs and could provide up to 1200 L/h of purified water from industrial effluent containing moderate total dissolved solids (TDS) and chemical oxygen demand (COD). NF process reduces all impurities in a single step including excess TDS, salinity, hardness, turbidity, heavy metals and microbial content. The operating cost in these plants is about Paise 3/L of purified water generated which reduces the volumetric load on evaporators employed for disposal of reject from NF process. The same NF setup is also used for decolorization of an aqueous solvent in textile industry due to the unique property of the NF membrane to allow monovalent salts to pass through but reject bivalent salts and larger organic molecules.

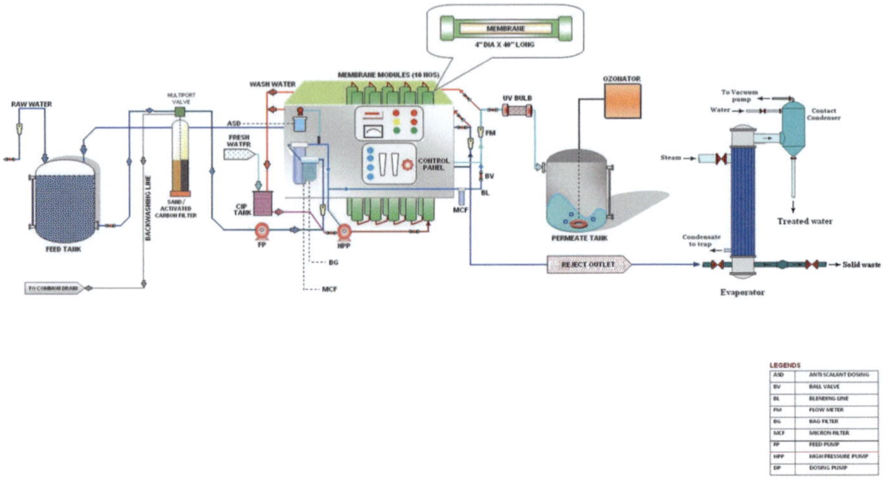

Fig. 3 Schematic of nanofiltration experimental setup to save evaporator energy consumption during effluent treatment or drinking water purification

Aqueous effluents from pharmaceutical industries comprise of solvents, salts and water. The possibility of recovering solvents and enable water reclamation is the objective of this study. One particular effluent from a pharmaceutical industry that produces antiretroviral drugs contains 15 % dimethylsulfoxide (DMSO) solvent (reaction medium) along with 2–3 of hazardous sodium azide (NaN_3) salt. The hazardous salt needs to be separated before the desalted liquor is subjected to distillation to recover both water and solvent.

1.1.3 Drinking Water Purification

Surface water contains TDS below the maximum permissible concentration of 500 ppm and hence does not require application of RO technology. However, this type of water contains a large concentration of suspended solids, turbidity and harmful microbes which cause water borne diseases such as jaundice, typhoid, cholera, gastroenteritis etc. Ultrafiltration (UF) is a low-pressure process useful for treatment of surface water including those from lakes, ponds and canals. Unlike RO process, UF does not separate any dissolved salts from water, but efficiently removes microbial content, turbidity, odor and color. Commercial UF membranes in the form of hollow fiber modules care being fabricated in several countries including India. The primary advantage is the low capital investment and operating cost due to the involvement of low pressures and high surface area per unit volume ratios (Fig. 4).

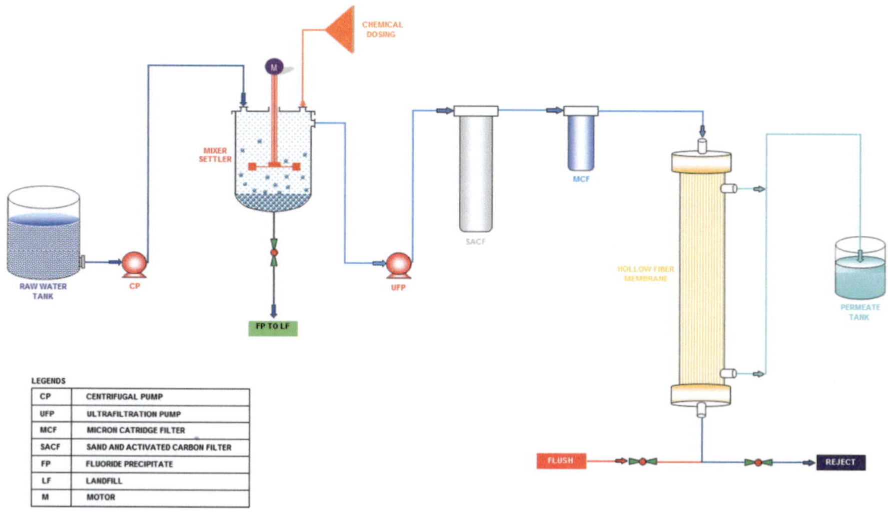

Fig. 4 Experimental set-up purification of surface water with low energy inputs

2 Result and Discussion

2.1 *Fuel Cells*

Ion exchange capacity (IEC) indicates the number of moles of exchangeable protons present per gram of polymer accountable for conduction as shown in Table 1. Study of methanol permeation through PEM is an important phenomenon due to its negative impact on DMFC efficiency. As shown in Table 1 coating of PANi on the surface of SPES reduces methanol permeability due to the formation of ionic bonds between anionic PANi and cationic SPES, which repels organic molecules but increasing attracts water. Proton conductivity of SPES/PANi polyion complex membrane is higher than bare SPES membrane as depicted in Table 1.

The emeraldine form of PANi contains numerous protonated imine nitrogen atoms which also contribute to proton conduction. The relation between temperature and proton conductivity of membrane is shown in Fig. 5a. Proton conductivity enhanced with rising temperature due to higher activation energy. Figure 5b depicts the Arrhenius plot used in the calculation of activation energy. The performance of

Table 1 Membrane properties at ambient temperature (25 ± 5 °C)

Membranes	Proton conductivity (S/cm)	Methanol permeability (cm^2/s)	IEC (meq/g)
PES	0.0026	3.7×10^{-7}	2.7
SPES	0.0317	9.25×10^{-8}	2.01
SPES/PANi	0.068	7.5×10^{-8}	3.7
Nafion 117	0.086	25.0×10^{-7}	0.91

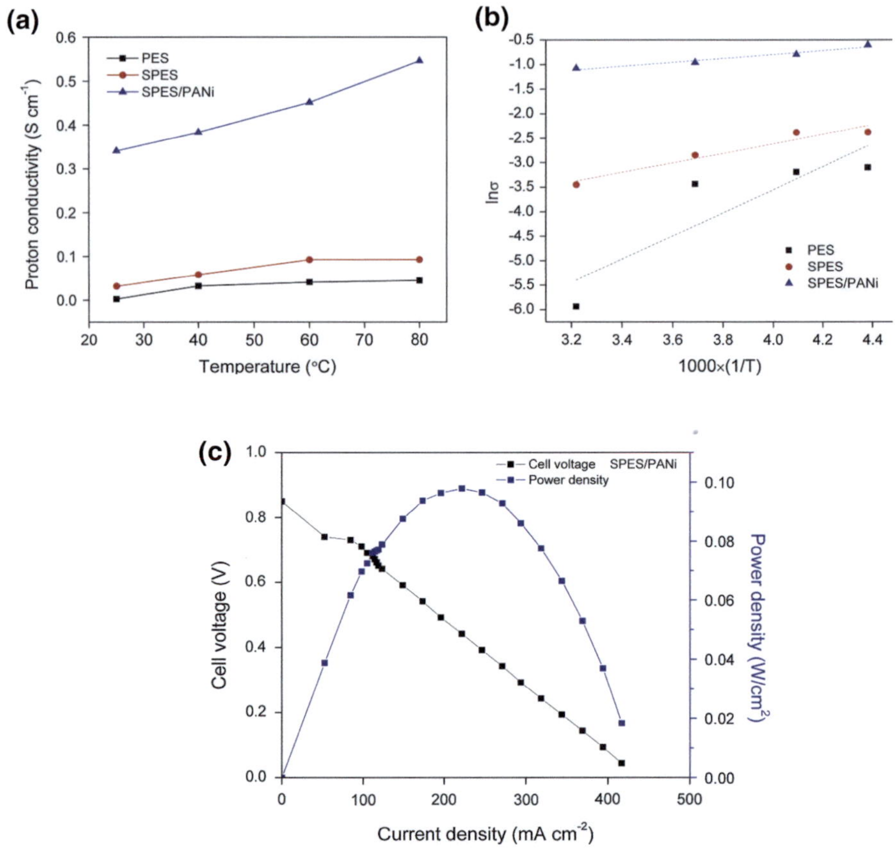

Fig. 5 Performance plots of PES/PANi polyion complex membrane. **a** Proton conductivity versus temperature. **b** Arrhenius plot. **c** Polarization curve

the developed membrane was evaluated at 25 ± 5 °C and atmospheric pressure as shown in Fig. 5c wherein a single slice fuel cell consisting of SPES/PANi membrane, gives a power density as high as 0.097 W cm^{-2} that corresponds to nearly 1 kW/sqm of membrane area used. The high energy density could be attributed to adequate water sorption, high IEC and proton conductivity besides low methanol permeability.

2.2 *Wastewater Treatment*

An electrodialysis-distillation integrated process was developed for the removal of the explosive sodium azide to enable recovery of DMSO solvent. Electrodialysis

(ED) was employed for removing the hazardous NaN_3 along with NH_4Cl from aqueous solutions without significantly changing the composition of the non-ionic constituents such as DMSO. The desalted feed was then sent to a two-stage distillation process for water removal and recovery of pure DMSO as shown in Fig. 6. The developed integrated process has been successfully designed and demonstrated to recover 25 kg of pure DMSO from a total desalted effluent quantity of 180 kg. The drug made with recycled DMSO solvent met all the desired market specifications of quality and purity.

Textile industries that produce acrylic fiber generate huge amounts of aqueous effluent stream containing 11 % thiocyanate (NaSCN) solvent along with 2–5 % of color imparting ions, salts and organic compounds such as β-sulfo propionic acid, β-sulfo propionitrile. Membrane based nanofiltration (NF) technique could be employed to separate 90 % of color imparting compounds and 85 % of total impurities from the industrial effluent. A five-stage NF operation with intermittent dilution of reject was found ideal to process 4000–6000 L/day of feed such that the combined permeate from first 3 stages could be sent to evaporator and spinning of the fiber whereas the permeate from last two stages could be reused in the next operation cycle as a diluent. The process showed more than 99 % NaSCN recovery and less than 0.17 % of NaSCN in final reject which made it fit for disposal as per the acceptable standards. The operating cost per m^3 of permeate generated was half of the conventional Gel Filtration Process that uses an expensive imported gel. Moreover NF offered a flexibility to enhance plant capacity for treating greater volumes of the effluent.

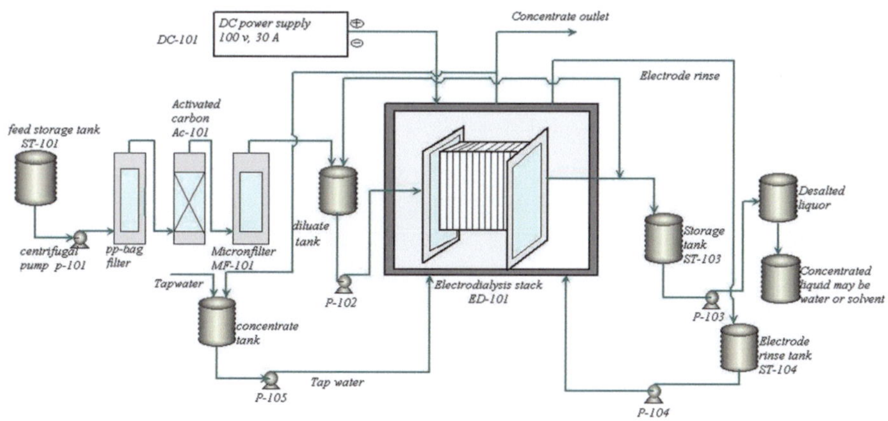

Fig. 6 Process flow diagram for desalination of aqueous effluent by electrodialysis to facilitate solvent recovery and enable energy savings in industrial ETP

2.3 Drinking Water Purification

Ultrafiltration modules of varying separation areas of 0.1–0.2 m^2 have been fabricated using hollow fibers made of polyacrylonitrile (PAN), polyethersulfone (PES), polyvinylidene (PVDF) and polyphenylsulfone (PPSU) polymers. The modules have been successfully tested for disinfection and clarification of surface water. PPSU membrane in particular was particularly robust and exhibited a flux of 100–200 L/m^2 h at an applied pressure of 2 atm with a turbidity rejection of 99.5 %, bacteria log reduction of 10^{-6} and water recovery of 90 %. UF systems based on hollow fiber membranes would be inexpensive by costing only Rs. 2.5 lakhs/- for producing 5000 L/h of purified water, due to the large membrane surface area per unit volume. Hollow fibers can be employed even without power wherever overhead tanks are available since a pressure of 0.5 atm is sufficient to force pure water through the membrane. This technology can be employed wherever surface water is available in plenty and operating cost would be below 3 paise/L (Fig. 7).

A hand operated system with submerged hollow fiber membrane has been designed for flood affected areas and remote villages which lack power supply. Further, funding and industrial support is required for taking this technology to the pilot plant level and future commercialization in regions where surface water is available in plenty. Figure 8 represents the hand operated ultrafiltration plant based on submerged spiral wound modules, designed for processing surface water, which

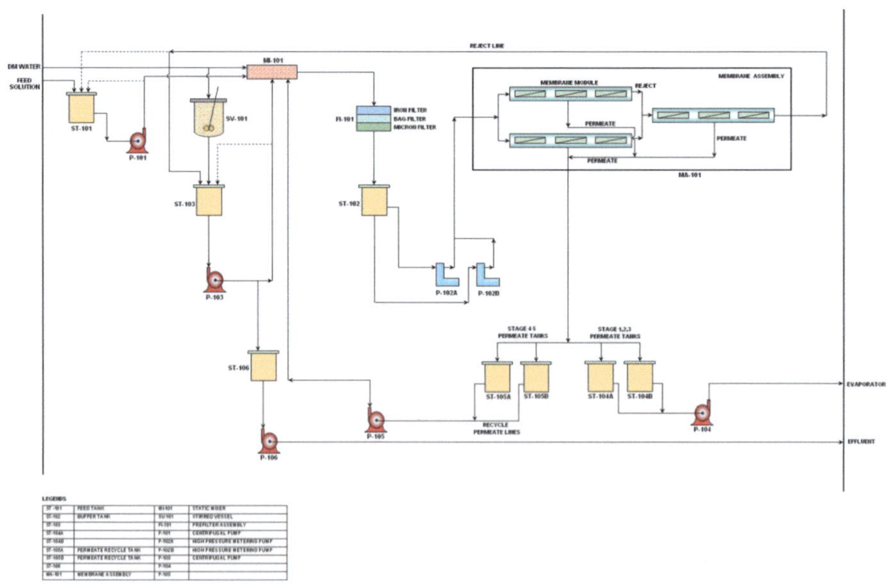

Fig. 7 Process flow diagram for energy efficient decolorization of solvent in textile industry

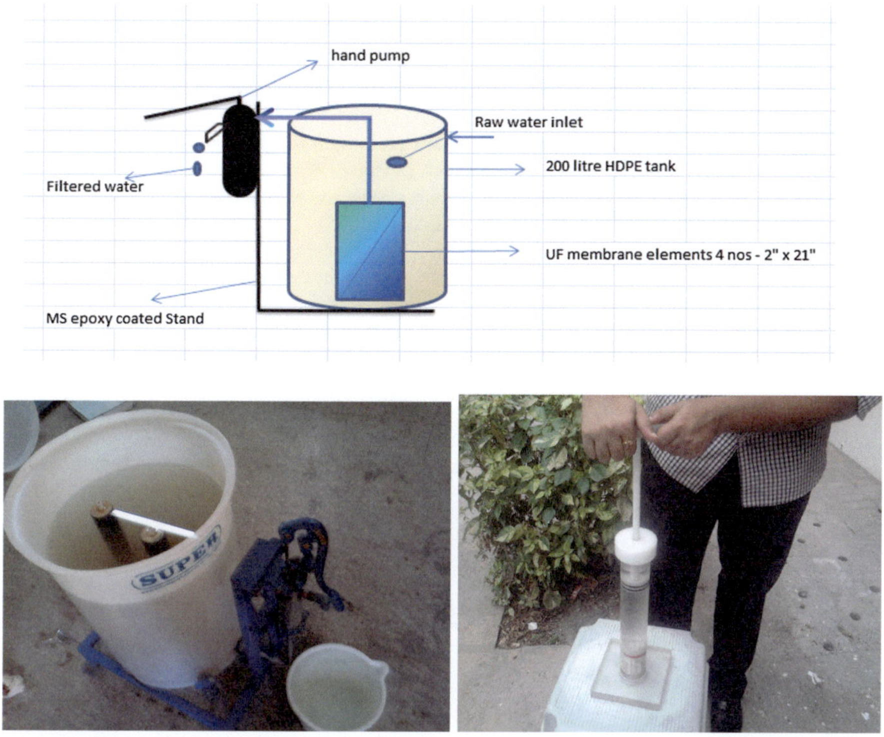

Fig. 8 Energy conservation through hand operated ultrafiltration for clarification and disinfection of flood water

has low TDS concentration, by removal of turbidity and microbial content to produce safe water at the rate of 250–500 L/h.

3 Conclusions

The sulfonated polyethersulfone (SPES) coated through oxidative polymerization of aniline membranes show significant proton conductivity and low methanol permeability making it useful for fuel cell application. The formation of hydrogen bonding due to the interaction between the sulfonated group and amine group in polyaniline, resulted enhanced conductivity with considerable compatibility. Therefore we can consider them as an attractive potential candidate for application in the field of fuel cells development. Membrane based integrated processes have an extensive role to play in wastewater treatment to enable solvent recovery and water reclamation. Hollow fiber membrane technology has exhibited tremendous scope for drinking water purification with minimum energy input.

References

1. Z. Zicheng, F. Yongzhu, M. Arumugam, Novel blend membranes based on acid-base interactions for fuel cells. Polymers **4**, 1627 (2012)
2. P. Boguslaw, Fuel cells—the future of electricity generation for portable applications. Environ. Biotechnol. **4**, 60 (2008)
3. L. Lei, W. Yuxin, Quaternized polyethersulfone Cardo anion exchange membranes for direct methanol alkaline fuel cells. J. Membr. Sci. **262**, 1 (2005)
4. C. Jinhua, A. Masaharu, M. Yasunari, Y. Masaru, Fuel cell performance of polyetheretherketone-based polymer electrolyte membranes prepared by a two-step grafting method. J. Membr. Sci. **319**, 1 (2008)
5. C. Dongju, Y. Shanshan, L. Xue, L. Xianfeng, Porous polybenzimidazole membranes with excellent chemical stability and ion conductivity for direct borohydride fuel cells. J. Power Sources **282**, 323 (2015)
6. Y.V.L. Ravikumar, S.V. Satyanarayana, S. Sridhar, Development of an electrodialysis-distillation integrated process for separation of hazardous sodium azide to recover valuable dmso solvent from pharmaceutical effluent. Sep. Purif. Tech. **110**, 20 (2013)
7. S. Sridhar, B. Smitha, New developments in nanofiltration technology: a case study on recovery of impurity-free sodium thiocyanate for acrylic fibre industry, ed. by A.K. Pabby, S.S. H. Rizvi, A.M. Sastre. Handbook of Membrane Separations: Chemical, Pharmaceutical, Food, and Biotechnological Applications, Chap. 42, pp. 1101–1129 (CRC Press, Taylor & Francis Group, Boca Raton, Fl, USA, 2008)
8. K. Praneeth, S.K. Bhargava, J. Tardio, S. Sridhar, Design of novel ultrafiltration systems based on robust polyphenylsulfone hollow fiber membranes for treatment of contaminated surface water. Chem. Eng. J. **248**, 297 (2014)

Transition to Lower Carbon Energy Regime: Engineering Challenges in Building and Transportation Sectors

K.V. Raghavan

Abstract Realising a smooth transition to lower carbon energy (LCE) in building and transportation sectors requires careful national level planning and multitasking with implementable milestones. This article examines the current and projected energy and GHG emission reduction scenarios, alternative energy sourcing options and the prevailing energy/emission rating and regulating systems. After a technology status review, certain impact making technology and engineering options suitable for implementation during the transition period have been identified for building and transportation sector.

Keywords Energy efficiency/intensity · GHG emission reduction · Thermal envelopes · Energy retrofits · HVAC systems · LDV/MDV/HDV · Hybrid propulsions · Light weighting

1 Introduction

If green house gas (GHG) emissions remain unchecked under business as usual scenario, the global average temperature is expected to increase by 6 °C by the end of 21st Century. It could have large scale humanitarian, environmental, political and socioeconomic repercussions. The building and transportation sectors together account for 50 % of global energy consumption and GHG emissions. Their transition to lower carbon energy (LCE) regime is very vital to mitigate and manage global climate. The technological transitions in the past were successful whenever the stakeholders received tangible benefits which made them voluntarily transform

K.V. Raghavan (✉)
INAE Distinguished Professor, Indian Institute of Chemical Technology, Hyderabad 500007, India
e-mail: kondapuramiict@gmail.com

K.V. Raghavan and P. Ghosh (eds.), *Energy Engineering*,
DOI 10.1007/978-981-10-3102-1_14

their life styles as demanded by the new situation. The transition process to LCE regime, therefore, needs to be customer friendly and carefully planned and executed at national level. An appropriate time horizon has to be fixed with implementable milestones. Various technological and engineering options have to be evaluated for their adoptability and the most sustainable ones to be selected for achieving energy efficiency enhancement and emission reduction in the utilization sectors. Novel Research, development and demonstration (RD&D) measures have to be launched through public-private partnership to accelerate their rate of technology absorption. The main task of the stakeholders in the building and transportation sectors is to create a congenial environment for generating attractive job and business opportunities so as to counter balance the additional cost of introducing LCE systems. From an engineering perspective, the transition process should have the dynamism to bring the necessary technological change(s) by maximizing the impacts of synergy, innovation and experience sharing to empower the stakeholders and promote entrepreneurship.

The International Council of the Academies of Engineering and Technological Sciences (CAETS) formed an Energy Committee in 2013 to study the global technological and engineering challenges associated with the transition to LCE regime in building and transportation sectors [1]. This presentation highlights its major findings and their implications on the growth of these sectors.

2 Energy and Emissions and Their Reduction Potential in Building Sector

The global building sector, comprising of residential and commercial segments, is worth USD 7.5 trillion accounting for 10 % of its GDP. The overall lifespan of buildings ranges from 40 to 70 years depending upon their type, quality, geographical location and maintenance. Their annual energy consumption in 2010 was reported to be 32.5 billion MWh (67 % residential) and is projected to reach 52.7 billion MWh by 2040 with an annual growth rate of 1.6 % [2] An estimated 0.8 billion people require new housing and 4.3 billion people have to be provided with modern energy resources. Among the emerging economies in Asia including China and India, Middle East and North Africa are expected to grow at greater than 3 % per annum. The energy end use in buildings [2] varies with climatic zone, socioeconomic status of people and technological advancements achieved (Fig. 1). The GHG emissions from the building sector in 2004 was placed at 8.6 Gt CO_2 eq. [3] with an average annual growth of 2 % and is projected to rise to 14.3 Gt CO_2 eq. by 2030 [4]. The countries with major GHG emissions in building sector are USA, China, EU-25, Russia, Australia and India. The emission abatement potential

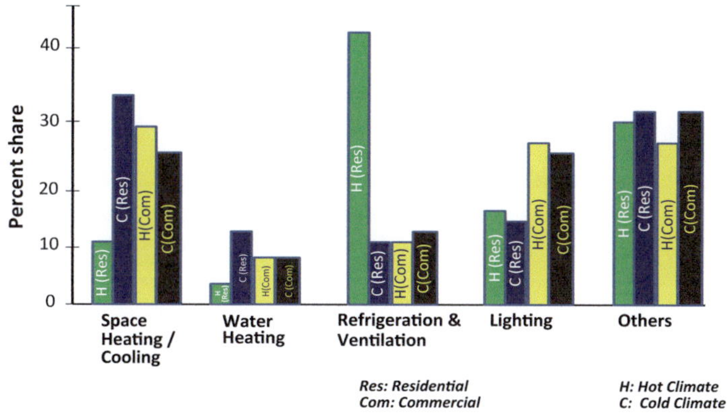

Fig. 1 Energy end use patterns in buildings

of global building sector is estimated to be around 3.52 Gt CO_2 by 2030 with 68 % accounted by the residential buildings [5]. Nearly 50 % of the contribution is expected through energy efficient building packages for new buildings and retrofits for existing buildings.

3 Alternative Energy Sourcing for Building and Transportation Sectors

Alternative energy is that which is produced or recovered without the undesirable consequences inherent in high carbon fossil fuel use. Key shifts to such alternative energies in building and transportation sectors will be pivotal in transitioning to lower carbon economy. The carbon intensity of fuels and various travel modes will be a guiding factor for alternative fuel selection. For e.g., the carbon intensity of natural gas is 469 as compared to 1001 Gt CO_2/KWh for coal and is a good choice for a short or medium term transition. Figure 2 highlights the primary and secondary fuel options for building and transportation sector. The increased penetration of natural gas, solar, biofuel and geothermal and solar-geothermal hybrid options is rather high in buildings. Though first generation biofuels are very attractive, the second generation biofuels based on lignocellulosic biomass, biomass to liquid fuels and biosynthetic gas have also good potential. The potential alternative energy sources for transportation sector are natural gas (CNG/LNG), biofuels (waste derived liquid biomethane and algal oils), hydrogen, electricity, solar, wind and nuclear energy.

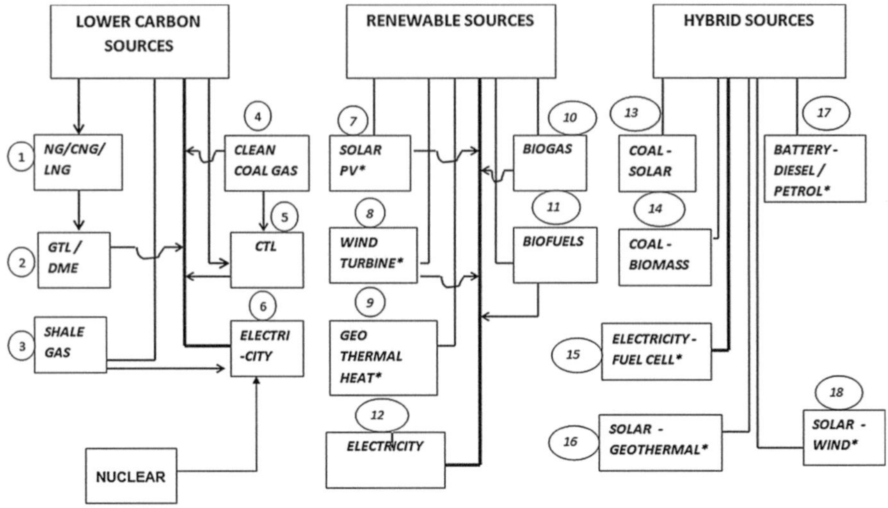

Fig. 2 Primary and secondary energy options for building and transportation sectors

4 Energy Classification, Ratings and Measurements in Buildings

The energy classification facilitates benchmarking of energy management in buildings. The passive energy concept prescribes building heating/cooling demand to be less than 10–20 kWh/m^2/year with air leakage not exceeding 0.6 times the building volume per year at 50 N/m^2. It facilitates to develop an integrated design approach for energy efficient space heating/cooling by optimizing its insulation, ventilation, air tightness, thermal bridges and heat recovery. By its adoption, the emission reduction can be as high as 27 % under global context by 2030. It is targeted to be achieved in residential buildings globally by 2030. The Net Zero Energy Buildings (NZEBs) produce as much energy as they use per unit time. The NZEB concept employs energy efficiency as first priority followed by the introduction of renewable energy onsite. It is targeted to be achieved in commercial buildings by 2030. The Net Positive Energy Buildings (NPEB) are based on an extended net zero energy concept to produce more energy than they consume.

For achieving transition to a lower carbon economy, a national energy rating system acts as a key instrument for improving the energy performance of new and existing buildings. The current regulations in building sector are predominantly GHG emission control driven. However, European Union's Energy Performance of Building Directive (EPBD), International Energy Conservation Code (IECC), American Society of Heating, Refrigeration and Air-conditioning Engineers (ASHRAE)-189, Energy Star, USA, Home Energy Rating System (HERS) of EU and National Australian Built Environment Rating System (NABRS) consider energy performance as the key rating factor. There is need to consider both energy efficiency and GHG emission reduction in an integrated fashion for building envelopes and HVAC units for devising appropriate technology and engineering interventions.

Measurement and Verification (M&V) are important tools for evaluating the energy and emission performance of buildings. A host of engineering tools are now being employed to perform online and offline measurements based on instrument physics, power electronics, flowmetry, thermography and image processing. More advances are anticipated in M&V engineering based on smart sensors, energy analytics and smart electric grids.

5 Technology and Engineering Options for Efficient Buildings

Two billion new buildings are likely to be constructed worldwide by 2050 with an estimated 255 billion square meters area at USD 3.7 trillion investment. Energy efficient buildings are expected to save annually 1.8 billion MWh and reduce GHG emissions by 0.5 Gt CO_2 eq. Since the life span of buildings being 40–70 years, their annual retrofitting rate is less than 2 % of the existing buildings resulting in an estimated 14 billion square meters of building space to be retrofitted by 2050.

From engineering perspective, the technological interventions are needed in developing energy efficient building envelopes and Heating, Ventilation, Airconditioning and Cooling (HVAC) systems and their retrofits and in devising advanced energy management systems. The envelope refers to the shell of a building which acts as a barrier to unwanted heat and mass transfer between its interior and exterior environments. Envelope design is a highly specialized engineering task with significant advances have been reported [6] in their spatio thermal modelling [7] green facading, insulation, air sealing, low conducting windows and special roof and wall structuring. Table 1 highlights the current and emerging technology and engineering options for energy and emission reduction in buildings. In recent years, the smart and solar envelope concepts [8, 9] are gaining thermo-physical prominence. They employ suspended particle devices for light/heat transmission through window glasses, phase change materials, algal facades for solar heat harvesting and antitoxin coated glasses. The solar envelopes reduce the cooling load by the shading effect and generate electricity on site [10, 11]. They are designed for 4 h operation in winter and 8 h operation in summer in cold countries.

The HVAC systems account for 30–35 % of building energy and 30 % of it can be reduced by employing energy efficient systems with the additional investment pay back of 3–5 years. From engineering perspective, heating/cooling, ventilation and air conditioning require maximum attention [12] with focus on heat pumps, combined heat and power (CHP) systems and onsite thermal storage (Table 1). Retrofitting of existing buildings and HVAC systems is quite challenging due to

Table 1 Current priorities and emerging options for efficient buildings

Item		Current priorities	Emerging technology options (2030)
1.1	• New buildings (ES: 1.8–2.5 ER: 0.4–0.5)	• Energy efficient building envelopes	• Passive (residential) and net zero energy (commercial) building targets • Smart and solar envelopes • Radiant chilled beam systems
		• Energy efficient HVAC systems	• Part Load performance as design basis • Heat pumping, thermostating and adsorption chilling • Displacement ventilation • LED and daylighting • High performance thermal storage • Hybridization of energy sources (renewable-nonrenewable)
		• Advanced energy management systems (large building complexes)	• Energy analytics with real time controls • Big data and cloud computing for intelligent systems • Networking with smart grids
1.2	• Existing buildings (ES: 3–6 ER: 0.35–0.7)	• Energy efficient envelope retrofits	• Thermal performance survey of existing buildings • Reflective surfacing • Closed cell spray insulation • Smart facade and window retrofits
		• Energy efficient HVAC retrofits	• Up gradation of coefficient of performance of buildings • Coupling solar thermal and photovoltaic systems • Variable speed drives • Micro gas turbines for combined heat and power systems (CHPs)

ES Energy Savings in billion MWh/Yr by 2050, *HVAC* Heating, Ventilation, Airconditioning and Cooling Systems, *ER* Emission Reduction in giga t CO_2/Yr by 2050

their extreme diversity in design basis, quality and maintenance. Pre-design structural, engineering and energy auditing is a major task prior for developing efficient shallow or deep retrofit options. Energy efficient building retrofits focus on minimization of air and water infiltration, mitigation of radiative thermal effects, introduction of thermal breaks and improvement of structural. The typical retrofit options for HVAC systems include coupling renewable energy to heating/cooling systems, variable speed devices, energy economized ventilation, comfort zoning and improved automation/control. The global investment on advanced Energy Management systems in poised to reach USD 511 billion by 2024. They consist of centralized, interlinked and networked systems [13] designed on the basis of energy analytics supported by UNIX, LINUX or WINDOWS and wireless technologies (Table 1). Advanced materials for energy efficient buildings has been receiving considerable attention in recent years (Table 2).

Table 2 Advanced materials for energy efficient building systems

No.	Building components	Application areas	Emerging material options
2.1	Building envelope	• Roof and walls	• Radiant barrier sheathing • Insulating concrete foams
		• Insulation	• Two component spray foams and nano structure polymeric foams • Vacuum insulation panels
		• Facade	• High reflectance coatings • Energy producing algae • Detoxicating titanium dioxide • Smart interactive materials
		• Windows	• Solar control and impact resisting glasses • Suspended particles for thermal/light control • Smart window materials
2.2	HVAC systems	• Heating/cooling	• Phase change materials for thermal storage • Thermoelectric materials
		• Ventilation	• Desiccants for energy recovery ventilation
		• Air conditioning	• Liquid desiccants (lithium salts/halide salt solutions)

6 Energy and Emission Reduction Potential in Transportation Sector

The World Economic forum (WEF) projected the energy consumption in 2030 under business as usual (BAU) and rapid drive scenarios (RDS) [14] for light and heavy duty road transport vehicles, rail, air, marine and pipeline transportation. The details are presented in Table 3 and they show that overall energy increase in transport sector can be anywhere between 12.5 and 40 % by 2030. The global stock of transportation vehicles is expected to nearly double during 2010–30 and the energy efficiency improvement potential is found to be maximum in the case of marine transport sector. The GHG emission potential of transportation sector is also projected under BAU and RDS scenarios in Table 3. The future is not sustainable under BAU scenario. Emission reduction potential of transportation sector can be improved through demand management as dictated by the consumer behavioural pattern. The transport modal choices differ from place to place which are caused by differences in socioeconomic conditions and urban development policies. Adoption of multimodal transport system also contributes to emission reduction. The McKinsey and Company has projected [15] the GHG emission abatement potential of transportation sector as 2.6 Gt CO_2 eq. per annum by 2030 with the cost of

Table 3 Global energy consumption scenarios for transport sector

Transport sector		Energy consumption (2010)		Projected energy consumption (2030)				
				BAU (CAGR: 1.7 %)		RDS (CAGR: 0.5 %)		EIP
		Billion MWh	% share	Billion MWh	% share	Billion MWh	% share	%
• Road	LDV	13.02	52	19.58	54.7	15.74	54.5	25
	HDV	5.47	21	6.07	16.9	5.57	19.3	15–30
	Sub-total	18.49	73	25.65	71.6	21.31	73.8	–
• Rail		0.76	3	1.02	2.9	0.66	2.3	15
• Air		2.55	10	4.72	13.2	3.61	12.5	30–40
• Marine		2.55	10	3.04	8.5	2.62	9.1	40
• Pipeline		0.76	3	–	–	–		10–20
• Others		0.26	1	1.35	3.8	0.66	2.3	NA
Total		25.37	100	35.78	100	28.86	100	

	GHG emissions (2010)			Projected GHG emissions (2030)			
				BAU		RDS	
	Gt CO_2 eq.	% share	t/t O eq	Gt CO_2 eq.	% share	Gt CO_2 eq.	% share
• Road	5.05	74.0	3.01	6.56	70.5	5.01	72.1
• Rail	0.08	1.2	1.21	0.09	1.0	0.08	1.2
• Air	0.75	10.9	3.41	1.24	13.3	0.70	11.8
• Marine	0.78	11.4	3.50	0.91	9.8	0.85	12.5
• Pipeline	0.17	2.5	2.5	0.50	5.4	0.16	2.4
	6.83	100.0		9.30	100.0	6.8	100.0

BAU Business as Usual Scenario; *EIP* Energy Improvement Potential; *RDS* Rapid Deployment Scenario
LDV Light Duty Vehicle; *MDV* Medium Duty Vehicle; *HDV* Heavy Duty Vehicle

abatement at Euro 60 per tonne of CO_2 eq. The rail (passenger and freight) and marine (freight) transport have the minimum GHG emission intensity followed by road transport. The air transport has the highest emission intensity. The light duty passenger vehicles have a wide band of GHG intensities due to variations in energy usage, fuel source and quality.

7 Energy/Emission Ratings for Transportation Sector

Road transportation is one sector in which regulations are highly effective in reducing GHG emissions. The EU, USA, China, India and other countries have set their own emission standards. The USA has also made regulations on energy efficiency in automobiles based on Corporate Average Fuel Economy (CAFE) concept keeping a target of achieving 35 mpg by 2020. The global fuel economy initiative was established recently [15] with a target to reduce the average fuel consumption of light duty vehicles to 50 % by 2030. Japan has introduced fuel economy regulations for heavy duty vehicles. Similar efforts have been reported for railway, air and marine transportation [16]. New regulatory framework for the transportation sector should be technology neutral, pro modal shift to encourage public transportation and fuel economy specific for each class of vehicle.

8 Technology and Engineering Options for the Energy Efficient Transportation

Multidisciplinary engineering skills of high order are required to develop energy efficient transportation systems with minimum GHG emissions. The CAETS Energy Committee has prioritized energy efficient road, rail, air and marine transport vehicle engines for technology and engineering interventions. The study conducted by the US-DOE in 2013 [17] shows that air and marine transport sectors have the maximum energy efficiency improvement and fuel decarbonisation prospects followed by road transport sector.

8.1 Road Transport

The energy efficiency improvement potential of light duty vehicles (LDVs) which constitute major part of road transportation is estimated to reach 35 % by 2034. Their fuel economy has already improved from 6 to 10 km per litre during 1978–2012. The electric LDVs in the form plug in hybrid (PHEV), battery electric vehicles (BEV) and H_2 fuel cell vehicles (HFCV) represent a leap forward in the

LDV energy efficiency. The present medium (MDVs) and heavy duty vehicles (HDVs) are predominantly diesel propelled with 45 % thermal efficiency. The first transition to LCE is achievable through a switchover to natural gas (CNG/LNG or shale gas) or biofuels or their diesel hybrids. Hybrid MDV/HDV power trains based on electric, hydraulic and flywheel options will receive increased attention in the coming years. Table 4 indicates the potential areas for present and future technology and engineering interventions for LDVs. They include electrification, advanced IC engines and advanced medium and heavy duty vehicle versions.

8.2 Rail Transport

The rail transport is far more energy efficient as compared to other transport modes. Track electrification and Hybrid diesel-electric locomotives offer real time energy efficiency benefits to countries with heavily diesel reliant rail system [18]. Significant technological advances have already been made in developing advanced engines, battery systems, regenerative braking and control units. Advanced reciprocating and turbocharged engines with fuel consumption less than 188 gallons per KWh, achieving smooth scalability with multiple engine systems, energy efficient traction propulsion retrofits, ultra super capacitors for brake energy storage and intelligent diagnostics/prognostics are emerging engineering options in this sector. The high speed rail sector need cutting edge power technologies, energy efficient static frequency converters and dry type transformers [19] Table 5 highlights them.

8.3 Air Transport

The heavily fossil fuel dependant aviation sector has been witnessing notable technological advances in recent years with better prospects for transitioning to lower carbon energy regime. The recent successful testing of biofuel and solar energy options for aero-engines are path breaking. The aero sector is always sensitive to engine upgradation efforts to achieve “higher and faster”. A shift to “greener and energy efficient philosophy” is taking place”. The current technological upgradation programmes of aero engines [20] are directed towards improved propulsion, thermal and aerodynamic efficiencies, multifuel compatibility, more intelligent control systems and light weighting of components (Table 6). The emerging technologies are more advanced open rotor engines, large eddy break up devices, hybrid laminar flow systems, innovative wing tip devices and control of boundary layer separation. Emerging options in control systems [21] include memory mapped microprocessor controlled ignition systems and advanced controls for monitory deteriorating engine performance [22]. More engineering breakthroughs are anticipated in evolving novel propulsion architecture to use cryogenic biofuels.

Table 4 Current priorities and emerging technology options for energy efficient road transport vehicle engines

S. No.	Item	Current priorities	Technology and engineering options	Emerging developments
1	Electrification of light duty vehicles (LDVs)	• Plug in electric vehicles (EVs) • Hybrid EVs (HEVs) • Plug in hybrid EVs (PHEVs) • Tesla Cars • Battery electric vehicles (BEVs) • H_2 fuel cell vehicles (HFCVs)	• Light weighting • Magnetic synchronous and hybrid motors • Level II chargers (conductive/inductive)	• Carbon fibre structurals • Mould finished chassis • Level III chargers • Asychronus induction AC motors
2	Advanced LDV-IC engines	• Advanced fuel injection and ignition • NO_x emission reduction • Adaptive control and after treatment • Thermal management • High speed direct injection diesel engines	• Variable compression ratio • Homogeneous charge compression ignition (HCCI) • Model based controls • Waste heat recovery • Light weighting	• Spark ignited direct injection turbo (SIDIT) • Atkinson cycle • Common rail diesel injection (CRDI) engines • Multivalve engines with variable timing and lift
3	Advanced medium and heavy duty vehicles (MDVs/HDVs)	• Diesel to CNG/LNG/biofuel • Low NO_x in emissions • High volumetric efficiency	• Direct injection/turbo charging • HCCI • High pressure direct injection • Multifuel injection	• Hybrid medium and heavy duty power trains • Aerodynamic drag reduction (20^+ %) • Low rolling resistance tyres
4	Advanced materials	• A practical trade off between engine efficiency, light weighting and fuel consumption • Improved aerodynamics • Performance improvement of vehicle components	• High performance polycarbonates/PU foams • Carbon composites • Surface engineering for higher wear resistance	• Advanced coatings/films • Novel composites and layered materials • Nanoscale inorganics

Table 5 Current priorities and emerging technological options for energy efficient rail transport

S. No.	Constituency/sub area	Current priorities	Technology and engineering options	Emerging developments
1	Electrification of railways	• Enhancing traction power network capacity • Kinetic energy recovery during breaking • Intelligent microelectronics and controls	• New engineering options • Regenerative breaking system • Insular gate bipolar technology (IGBT) • Intelligent rail electronics	• Power quality conditioners • Ultra super capacitors for brake energy storage • Magnetic levitation technology
2	Hybrid diesel electric locomotives	• Fuel flexibility and diesel use reduction • Hybrid traction propulsion • Improved engine performance	• Hybridization of Energy resources • Regenerative braking system • Multiple diesel engines and three phase induction motors	• Fuel efficiency of 24^{+} Kms/l of diesel • Modern speed controls for induction motors • Hybrid traction propulsion retrofits
3	Advanced version diesel engines	• Advanced reciprocating engine technology • Higher efficiency and more reliability • Fuel Consumption: <188 gallons/kWh	• 4 stroke turbo charged engine with temperature control • Scalability with multiple engine systems	• Advanced diagnostics and prognostics • Static frequency converters • Dry type transformers • Adaptability to biofuels • Use of metal composites

Table 6 Current priorities and emerging technology options for energy efficient aero engines

S. No.	Item	Current priorities	Technology and engineering options	Emerging developments
1	Advanced aero engines	• Greener and energy efficient with 50 % CO_2 reduction by 2020 • Improved propulsion, thermal and aerodynamic efficiencies • Model based controls	• Advanced turbo fan and open rotor architecture; blended wing bodies • Friction and lift drag reduction • Natural or hybrid laminar flow for minimization of wetted areas	• Multifuel compatibility • Achieving thermal efficiency threshold of 70 % • Improved thermal management and power offtakes • Memory mapped microcompressor based ignition control
2	Advanced materials	• Enhanced temperature capability • Adaptive materials for reduced fuel burn/emissions • Light weighting	• Composites with non metallic fibres in polymer matrix • Aluminium-lithium alloys • Dual microstructured nickel alloys • Smart materials embedded with nano particle	• Recyclable composites; ceramic matrix composites • Titanium alloys for 650^{+} °C service • Single crystal alloys • Self monitoring and smart material performance enriched by nano technology
3	Fuel decarbonization efforts	• Biofuel • Solar Energy	• Successful engine tests of biofuels • Long distance solar powered flight	• 2 million tpa biofuel by 2020 for aviation • Solar powered energy packs (ground)

8.4 *Marine Transport*

The diesel propelled Engines are the main workhorses of marine transport [23]. In recent years, they are upgraded through the introduction of turbo charging, adjustable cam shafts, variable inlet valve control, improved combustion and exhaust gas recirculation. The emerging technologies are expected to make them more intelligent [24] with focus on gas turbine marine engines with isentropic compression/expansion and isobaric combustion run on Brayton cycle, hybrid marine engines (solar/wind/biodiesel) and electric propelled engines with super conducting DC homopolar motors. Long term prospects are foreseen for nuclear propulsion for super tankers and large container ships (Table 7).

Table 7 Current priorities and emerging technology options for energy efficient marine engines

S. No.	Item	Current priorities	Technology and engineering options	Anticipated/emerging developments
1	Advanced marine engines	• NO_x and SO_x reduction • Enhanced energy efficiency • Multifuel compatibility	• Variable inlet valve control; turbo charging • Gas turbine engines • Hybrid engines	• Intelligent engine drives • Magneto hydrodynamic propellers • Super conducting DC homopolar motors
2	Advanced materials development	• Higher resistance to marine environment • In situ reclaimed materials for rail road beds • Smart operation Facilitation • Special coatings for surface protection	• Special ferritic steels • Light weight plastics • Nickel and cobalt based super alloys • Structural adhesives	• Advanced materials for ship super structures • Materials for shape changing hulls • Advanced underwater coatings to prevent biofouling of ship bottoms
3	Fuel decarbonization options	• Shift from diesel • Renewable energy for onboard facilities • Efficient propulsion for jumbo cargo vessel	• LNG/Biofuel • Solar/wind/fuel cells for auxiliary power units • Nuclear propulsion (long horizon)	• Solar powered small ships • Onboard power packs • Hydrogen as fuel (long horizon)

9 New Materials for Transportation Sector

The light weighting of vehicle engines and bodies has become a vital factor for achieving higher energy efficiency in all transport systems (Tables 4, 5, 6 and 7). In road transport vehicles, new light weight materials are employed for suspension, brake fittings, wheels and vehicle bodies. High specific strength, light weight and enhanced thermal capability are favoured in aero space applications [25] with composites, glass, carbon, kevlar, ceramics and polymers as important materials. Smart materials are likely to be used in future for aircraft wings and exhaust nozzles and also nanomaterial incorporated materials for achieving self healing and self monitoring capabilities. In case of marine transport systems, high strength structural steels (ferritic), light weight plastics, nickel and cobalt based super alloys and smart materials for vibration damping, thermal regulation and shape changing hulls are most favoured new materials.

10 Summary

The transition to low carbon energy regime in building and transportation sectors is a dynamic process of technological change propelled by innovation, synergy, experience sharing, empowerment and entrepreneurship. A wide range of technology and engineering options are available for both sectors. The priority areas for building sector are new and retrofit building envelopes and HVAC systems and advanced energy management tools. The need for achieving passive and net zero energy standards for residential and commercial buildings respectively by 2030 is a major milestone to be achieved in the transition process.

The road, rail, air and marine transportation heavily rely on fossil fuels. There are definite indications of their moving towards a lower carbon energy regime. The priority areas are energy efficient LDV/MDV/HDV, hybrid electric-diesel locomotives, advanced aero and marine engines with partial fuel decarbonisation and switchover to multimodal transport options. Advanced materials play a vital role in the light weighting with higher strength and high reliability.

Acknowledgments The speaker is grateful to Dr Baldev Raj for his valuable contribution as the co-editor of CAETS Report and as President, CAETS and the intellectual contributions from the members of the CAETS Energy Committee viz., Dr Philip Lloyd (South Africa), Dr J. Loughhead (UK), Dr Vaughan Beck (Australia), Dr R. Evans (Canada), Prof F Behrendt (Germany), Dr R Huegli (Switzerland), Prof M Ihara (Japan) and Dr Peng Suping (China). The speaker acknowledge the valuable support provided by Dr B N Suresh, President, INAE and is thankful to Mr R R Yadu Krishna of IICT for assistance in document preparation.

References

1. CAETS, in *Transitioning to Lower Carbon Economy: Technological and Engineering considerations in Building and Transportation Sectors*, ed. by K.V. Raghavan, A. Baldev Raj. Report from CAETS Energy Committee (2015)
2. O. Lucan, D.U. Verstz, in *CC Buildings: Climate change 2014*, Chapter 9 of the 5th Assessment Report of IPCC, Cambridge, UK and USA (2014)
3. M.D. Levine et al., in *Residential and Commercial Buildings*, 4th Assessment Report of IPCC, Cambridge, UK (2007)
4. Grantham Institute for Climate Change, in *Reduction of CO_2 Emissions in the Global Building Sector to 2050,* Report GR3 from Imperial College, London (2011)
5. McKinsey and Co, in *Pathways to a Lower Carbon Economy*, A Report from McKinsey and Co (2009)
6. S. Asif, *Advanced building technologies for sustainability* (John Wiley & Sons Inc., 2012), 115 pages

7. A. Aksamija. Building simulations and high performance buildings research: use of building information modeling (BIM) for integrated design and analysis. Perkins + Will Res. J. **5**(01), 20–34 (2013)
8. M. Beevor, *Smart Building Envelopes* (A Report from University of Cambridge, UK, June, 2010)
9. A.A. Chernousov, B.Y.B. Chan, Thermal characterization of smart and large scale building envelope system in a subtropical climate. Int. J. Civil, Struct. Constr. Architectural Eng. **9**(4), 336–340 (2015)
10. R.L. Knowlers, The solar envelope: its meaning for energy and buildings. Energy Build. **35**, 15–25 (2003)
11. A. Vartholomais, The residential solar block envelope: a method for enabling the development of compact urban block with high passive solar potential. Energy Build. **99**, 302–312 (2015)
12. Principles of Heating, In *Ventilation and Airconditioning*, *ASHRAE Handbook—Fundamentals*, 7th edn. (2013), 600 pages
13. S.C.M. Hui, Latest trends in building automation and control systems. Paper presented at CAI symposium on intelligent facility management and intelligent transport, Hong Kong, 28th March 2007
14. WEF, in *Repowering Transport*, Project White Paper and Cross Industry Report, Geneva, April (2011)
15. R. Smokers, H.V. Essen, B. Kampman, E.D. Boer, R. Sharpe, in *EU Transport GHG: Routes to 2050*, A Report on Regulations for vehicles and Energy Carries by AEA/CE/TNO/ISIS, 12 February (2010)
16. UNFCCC, in *Methodological Issues under the Convention: Emissions from Fuel used for International Aviation and Maritime Transport*, Agenda 10(D) of 36th session of Subsidiary Body for Scientific and Technical Advice, Bonn, 14–25 May (2012)
17. Transportation Energy Future Series, *Commercial Trucks, Aviation, Marine Modes, Railroads and Pipelines and off Road Equipments* (A study by the Argonne National Laboratory, USA-DOE, 2013)
18. S. Frey, in *Railway Electrification Systems and Engineering* (A Book from Whiteword Publications, Delhi, India) (2012) ISBN 978-81-323-4395-0
19. ABB, *Powering the World's High Speed Rail Networks* (2014). www.abb.com/railway
20. R. Avellan, in *The Design of Energy Efficient Aero Engines: Some Recent Innovations*, Report with ISBN 978-91-7385-564-8 from Chalmers University of Technology, Gothenburg, Sweden (2011)
21. S. Martin, I. Wallace, D.G. Bates, in *Development and Validation of Aeroengine Simulation Models for Advanced Control Designs*, American control Conference, Seattle, USA, 11–13 June (2008)
22. Z. Weili, in *Performance Deterioration Evaluation of Aircraft Engines based on Simulation*, Prognostics and System Health Management Conference, China, 24–27 August (2014)
23. RA. Eng, in *Future Ship Powering Options: Exploring Alternative Methods of Ship Propulsion*, Report from Expert Working Group of RAE, UK, July (2013)
24. Frescobot, *Intelligent Diesel Engines.* www.en.wikiepedia.org/wiki/intelligent-Diesel-Engine. 20 January (2015)
25. McKinsey & company, *Lightweight, heavy Impact.* www.mckinsey.com/^/media/mckinsey/datcom/uploads/advanced-aircraft (2008)

Energy Options and Scenarios for Transitioning to a Lower Carbon Economy: An Indian Perspective

Ajit V. Sapre

Abstract Many lower carbon ways of producing energy and hydrocarbons are being developed around the world. While the contribution from solar and wind energy is rapidly increasing, biomass routes are another promising option to capture the Sun's energy. This article describes some bio-based technologies for the production of fuels and chemicals that may be attractive for agrarian economies like India. These technologies utilize surplus agri-residue, oils derived from plants such as jatropha and pongamia, and algae. In addition to environmental benefits, these options can provide energy and hydrocarbon raw material security, lower forex outgo, while creating employment and income opportunities for the rural population.

Keywords India · Algae · Jatropha · Agri-residue · Bio-ethanol · Bio-diesel

1 Introduction

India is currently the third largest energy consumer in the world. India has 0.8 % of the world's gas reserves, 0.3 % of the world's oil reserves and 6.8 % of the world's coal reserves [1]. India imports approximately 77 and 38 % of its crude oil and natural gas requirements respectively. Even at relatively lower world market prices of $46/bbl for crude oil and $2.7/MMBTU for natural gas, these imports represent more than $50 billion of annual revenue outgo. In fact, the landed costs and actual revenue outgo are quite higher because of the costs involved with transporting these hydrocarbons to India. This trend is likely to increase as India is expected to enter a more energy intensive phase of its development (Table 1, data obtained from the World Bank [2]).

The current power scenario in India is shown in Fig. 1. The data for this figure was obtained from the Central Electricity Authority of India [3]. A comparison of

A.V. Sapre (✉)
Reliance Industries Limited, Mumbai, India
e-mail: ajit.sapre@ril.com

K.V. Raghavan and P. Ghosh (eds.), *Energy Engineering*,
DOI 10.1007/978-981-10-3102-1_15

Table 1 India is likely to enter an energy intensive phase of development

Country	Per capita CO_2 emissions (metric tons, 2011)
USA	17
Russia	12.6
Japan	9.3
UK	7.1
China	6.7
India	***1.7***
World average	***5***

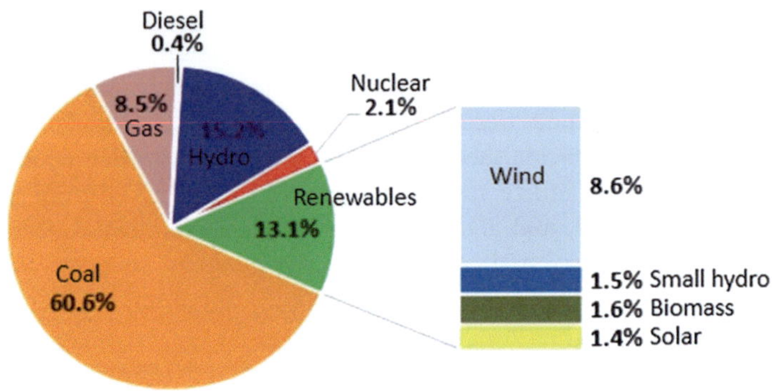

Fig. 1 India's power generation capacity by source (2014–15)

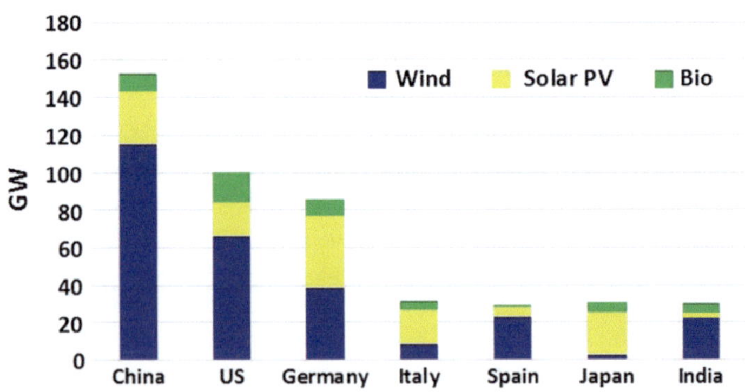

Fig. 2 Major renewable installed capacity in select countries excluding hydropower, 2014

India's renewable installed capacity with other countries is shown in Fig. 2, the data for which was obtained from REN21 [4]. These figures show that there is significant potential to increase the contribution from renewable to the energy mix in India.

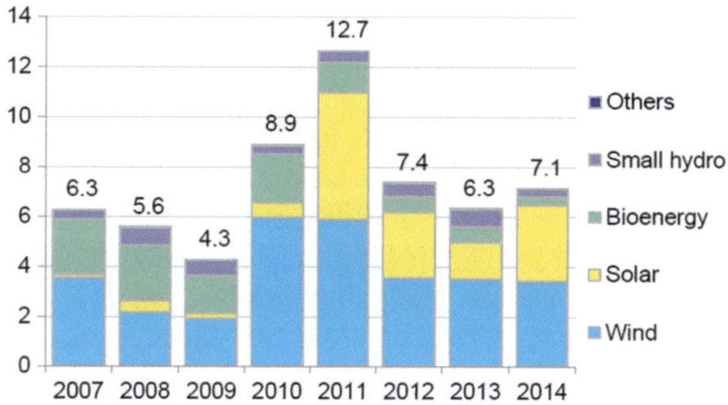

Fig. 3 Renewable energy investments in India (billion $) [5]

While there is a growing emphasis on solar energy investments in India in recent years, investments in bioenergy are lagging behind (Fig. 3, [5]) in spite of unique benefits such as the production of energy in easily storable and transportable form (hydrocarbons) and the creation of rural employment opportunities. This article describes some bio-based technologies that may be particularly attractive for India.

2 Bio-based Routes to Fuels and Chemicals

In order to improve the likelihood of commercial success, bio-based technologies should focus on resources (land, feedstock, water) with otherwise limited uses or no uses. They should also preferably not depend on subsidies or fossil fuel penalties for long term viability. Some examples are discussed below.

2.1 Surplus Agri-residue to Hydrocarbons

Burning waste agri-residue on fertile land can quickly clear fields, but this practice leads to loss of nutrients and causes significant pollution. Surplus agri-residue can instead be collected and converted into liquids such as kerosene at the village level, thereby improving the quality of life by addressing lighting and cooking needs.

Conversion of 1 kg of biomass with an energy content of approximately 4,000 kcal to power in a combined cycle biomass power plant can yield 1,465 kcal of energy. Alternatively, conversion of 1 kg of biomass to ethanol by enzymatic hydrolysis and fermentation can yield approximately 0.2–0.25 kg ethanol i.e. 1,420–1,780 kcal of energy. Ethanol yield from Indian non-feed/fodder biomass can be lower than from other highly cellulosic biomass due to their recalcitrant nature. Conversion of 1 kg of biomass to hydrocarbon fuel by catalytic pyrolysis and hydrotreatment can yield approximately 0.28 kg high energy fuel i.e. 2,960 kcal of energy. Furthermore, such hydrocarbon products are similar to crude derived products leveraging existing infrastructure investments. These estimates show that conversion of biomass to fungible hydrocarbons like kerosene is more energy efficient. The thermochemical platform to make hydrocarbons is also generally more carbon efficient than a biochemical (sugar) platform. Moreover, the thermochemical platform is invariant to the variety of non-fuel/fodder biomass.

This technology can replace traditional centralized power generation followed by distribution over long distances. India has estimated surplus biomass availability of 120–150 MMTPA [6]. This can potentially contribute more than 30 MMTPA of useful hydrocarbon fuel. Furthermore, agri-residue aggregation and supply can create tremendous employment and income opportunities for the rural economy. Challenges for commercial scale implementation of this technology would involve efficient handling of biomass, a short window of harvesting, moisture, grits, storage, capacity utilization, etc.

2.2 *Jatropha Based Biodiesel*

The Government of India has proposed an indicative target of 20 % for blending of biodiesel by 2017 [7]. This represents a very aggressive biodiesel growth trajectory, as biodiesel production in India in 2014–15 was only 0.1 MMTPA [8] corresponding to a blending level of less than 0.2 %. A major constraint for higher biodiesel production has been the lack of availability of desired feedstock; used cooking oils and residue from vegetable oil refineries are currently the main feedstocks [8]. Crops such as jatropha and pongamia grow efficiently on low rainfall marginal lands and can help address the feedstock issue. A key challenge is enhancement of crop productivity. Subsequent extraction and fuel conversion technologies are proven and simple. The crops can be subjected to hybridization and varietal improvement for enhanced virus tolerance, year-round flowering, improved branching patterns, increased yield and oil content, adaptability to different soils and climates, etc. For instance, such modern biotechnology techniques have significantly improved jatropha yields from <1,000 kg seeds/ha for wild jatropha to 3,000–4,000 kg seeds/ha for composite and hybrid varieties to 7,000 kg seeds/ha for top hybrids, while also greatly accelerating the time taken by these plants to reach production maturity [9]. Plantation of such crops can support India's biodiesel ambitions while creating rural employment opportunities.

2.3 *Algae to Ethanol and Hydrocarbons*

Bio-ethanol blending is being mandated by an increasing number of countries. The Government of India has also proposed an indicative target of 20 % for blending of bio-ethanol by 2017 [8]. Ethanol production in India is mainly from molasses—a byproduct during production of sugar from sugarcane. India was the second largest producer of sugarcane in the world in 2013. Brazil, the global leader in sugarcane

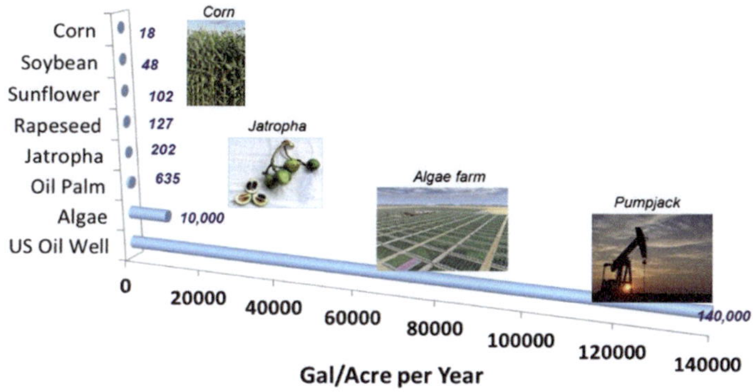

Fig. 4 Areal productivity of various feedstock resources

production, has some units that often exclusively produce ethanol (no sugar) using not only molasses but also sugarcane juice. Most US ethanol production is from corn. The mandate-driven global market for 1st generation corn ethanol is beginning to get saturated while 2nd generation cellulosic ethanol is currently challenged by high feedstock and capital cost. These bioethanol technologies have low productivity—less than 1,000 total gallons of liquid fuel (TGOLF) per acre-year.

As an alternative, algae are efficient converters of sunlight to hydrocarbons as shown in Fig. 4. Macro and micro algae being developed by leading companies use salt water, CO_2 and sunlight and can yield hydrocarbon fuels, ethanol and biochemicals with high photon to carbon fixation efficiencies. Algae derived jet fuel is also looking promising. Algae based technology will not compete with farm land or food crops and can in fact also deliver some fresh water as a byproduct. Engineering innovations in open ponds, closed photo bioreactors and downstream processing continue to occur. Advanced biotechnology tools such as metabolomics, genomics, protein science, systems biology, automated screening, etc. can improve algae yield, robustness and stress tolerance. Successful commercial deployment of such a 3rd generation ethanol and hydrocarbon technology can convert a coastal barren area into a field with never-declining productivity.

3 Global Outlook for a Bio-based Low Carbon Economy

There is a strongly growing nexus between rapidly advancing synthetic biology and engineering. We are already in the midst of an evolving bio-based low carbon economy. Commercial production of 1st generation corn ethanol, 1st generation biodiesel and 2nd generation cellulosic ethanol was 24.6 billion gal, 19.9 billion gal and less than 0.1 billion gal respectively in 2014. Corn ethanol production is eventually expected to plateau as mandates get met. Currently, most biodiesel production is from palm oil, soybean oil and rapeseed oil. Contributions from these and other crops such as jatropha and pongamia are expected to increase especially in countries that have significant dependence on diesel. For cellulosic ethanol, significant capacity is coming online quickly but the technology is challenged by feedstock cost. No commercial production from 3rd generation technologies such as algae was reported in 2014. Customers have very low appetite for paying premiums for green products. Breakthrough innovations are thus needed for commercial viability. Bio-based lower carbon technologies have also gradually started shifting their focus from fuels to higher value chemicals with the recent decline in crude oil prices.

4 Looking Forward

With increasing affordability of renewable electricity, a hydrogen-based or methanol-based economy could dominate in the future. Commercial hydrogen fuel cell vehicles are already available and sales are reported to be increasing [10].

Compared to developed nations, India's population has significant diversity in affluence. Figure 5 shows that for developed nations, additional economic growth requires little additional energy relative to developing nations. While India could follow the same growth trajectory as other developed countries, India should instead pioneer a more sustainable growth trajectory. Some of the technologies outlined in this article can help India not only bridge the gap, but leapfrog into the future.

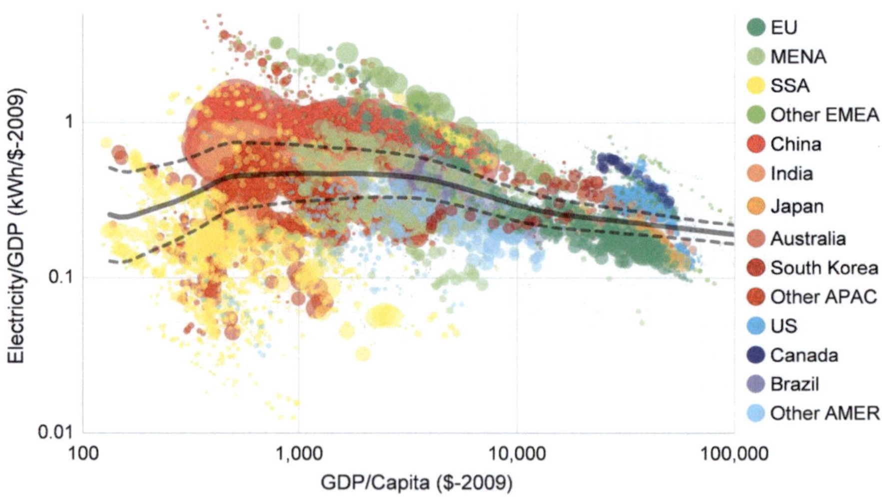

Fig. 5 Electricity consumption per unit of GDP versus GDP per capita, 1980–2012 [11]

References

1. BP Statistical Review of World Energy June 2015. http://www.bp.com/en/global/corporate/about-bp/energy-economics/statistical-review-of-world-energy.html
2. CO_2 emissions (metric tons per capita), *The World Bank*. http://data.worldbank.org/indicator/EN.ATM.CO2E.PC
3. Growth of Electricity Sector in India from 1947–2015. http://cea.nic.in/reports/planning/dmlf/growth_2015.pdf
4. Renewables 2015 Global Status Report. http://www.ren21.net/status-of-renewables/global-status-report/
5. B. Bhushan, *H1 2015 India Market Outlook: Groundwork for a New Era*, Bloomberg New Energy Finance, Jan 2015, reproduced with permission
6. Ministry of New and Renewable Energy (MNRE), Government of India. http://mnre.gov.in/schemes/grid-connected/biomass-powercogen/
7. National Policy on Biofuels, *Ministry of New and Renewable Energy, Government of India*. http://mnre.gov.in/file-manager/UserFiles/biofuel_policy.pdf
8. Ethanol Blended Petrol & Bio Diesel Policy—Seventh Report, *Standing Committee on Petroleum & Natural Gas* (2014–15), Sixteenth Lok Sabha. http://164.100.47.134/lsscommittee/Petroleum%20&%20Natural%20Gas/16_Petroleum_And_Natural_Gas_7.pdf
9. Future Harvest—21st Century Jatropha, *Hardman Agribusiness*, March 2015, http://www.hardmanagribusiness.com/future-harvest-21st-century-jatropha/
10. A. Rybczynska, *Q3 2015 Global Electrified Transport Market Outlook*, Bloomberg New Energy Finance, August 2015
11. N. Bullard, *Sparklines 41: Beautiful Evidence of Electricity's Future*, Bloomberg New Energy Finance, June 2015, reproduced with permission

Mixed Culture Chain Elongation (MCCE)—A Novel Biotechnology for Renewable Biochemical Production from Organic Residual Streams

W.S. Chen, M. Roghair, D. Triana Mecerreyes, D.P.B.T.B. Strik, Carolien Kroeze and C.J.N. Buisman

Abstract MCCE is a novel biotechnology that has potential to produce biochemicals from organic residual streams in a clean, renewable and economically viable way. A pilot plant has been established by ChainCraft in Amsterdam, Netherlands to process supermarket waste into value added biochemicals. Ongoing and future researches will aim to optimise the technological and environmental performances through Life Cycle Assessments (LCA) and alternative substrates to ethanol.

Keywords Organic waste · Biochemicals · Mixed culture biotechnology · Biorefinery

1 Introduction

Mixed Culture Chain Elongation (MCCE) is a novel biotechnology developed by Sub-department of Environmental Technology (ETE) to convert organic residual streams into biochemicals, i.e. Medium Chain Fatty Acid (MCFA; Saturated fatty acid with 5–9 carbons) using mixed culture microorganisms [1]. It consists of two steps: First, acidification in which complex organic matters in the residual streams are bio-degraded into basic building blocks like acetate and CO_2. Second, a chain elongation step in which acetate and CO_2 is elongated with externally added ethanol

W.S. Chen · M. Roghair · D. Triana Mecerreyes · D.P.B.T.B. Strik (✉)
C.J.N. Buisman
Sub-Department of Environmental Technology (ETE),
Wageningen University, Wageningen, The Netherlands
e-mail: david.strik@wur.nl

W.S. Chen
e-mail: wei-shan.chen@wur.nl

W.S. Chen · D. Triana Mecerreyes · C. Kroeze
Environmental Systems Analysis Group (ESA), Wageningen University,
Wageningen, The Netherlands

K.V. Raghavan and P. Ghosh (eds.), *Energy Engineering*,
DOI 10.1007/978-981-10-3102-1_16

into MCFA for diverse biochemical applications. MCCE has many advantages: (I). It is a simple biotechnology as sterilisation and genetic modification are not needed; (II). The broad feedstock spectrum includes various low grade biomass residual streams; (III) The product separation is easier as the products have longer carbon chain and thus lower solubility in water. These characteristics make MCCE a potentially clean, renewable and economically viable bioprocess.

2 Result, State-of-Art and Conclusion

ETE has continued enhancing and promoting MCCE from various perspectives: Industry-wise an MCCE pilot (ChainCraft in Amsterdam) has been implemented to process the supermarket food waste and ethanol into MCFAs. Environmentally an LCA is being applied to MCCE to evaluate its environmental performance and to design optimisation strategies. Technologically, we attempt to maximise the efficiency of ethanol use and to substitute ethanol with alternative substrates. We performed long-term MCCE operation to gain in-depth understanding of the conversion kinetics, upon which we can stimulate MCCE towards desired and efficient production [2]. Moreover, alternative substrates including electrons [3], H_2, syngas (H_2, CO and CO_2) and methanol [4] are tested as potential substitute to ethanol. The tested alternative substrates can be produced from biomass residues and are potentially cheaper, easier accessible and more renewable. Overall, we aim at continually enhancing MCCE into a clean, renewable, economically viable and geographically unbound biorefining process producing useful chemicals from organic waste and biomass residue.

References

1. K.J.J. Steinbusch, C.M. Plugge, H.V.M. Hamelers, C.J.N. Buisman, Biological formation of caproate and caprylate from acetate: Fuel and chemical production from low grade biomass. Energy Environ. Sci. **4\011**, 216
2. T.I.M. Grootscholten, K.J.J. Steinbusch, H.V.M. Hamelers, C.J.N. Buisman, Improving medium chain fatty acid productivity using chain elongation by reducing the hydraulic retention time in an upflow anaerobic filter. Bioresour. Technol. **136**, 735 (2013)
3. M.C.A.A. Van Eerten-Jansen, A. Ter Heijne, T.I.M. Grootscholten, K.J.J. Steinbusch, T.H.J.A. Sleutels, H.V.M. Hamelers, C.J.N. Buisman, Bioelectrochemical production of caproate and caprylate from acetate by mixed cultures. ACS Sustain. Chem. Eng. **1**, 513 (2013)
4. W.S Chen, Y. Ye, D.P.B.T.B. Strik, C.J.N. Buisman, Methanol as an alternative electron donor in Mixed Culture Chain Elongation (MCCE) for butyrate and caproate formation. Biomass Bioenergy, Manuscript Submitted for Publication (2015)

Bio-energy and Bio-refinery: Advances and Applications

Akshat Tanksale

Abstract According to current projections, global crude oil reserves will run out in less than 50 years (Source: IEA). A viable alternative to crude oil is urgently needed. Lignocellulose is the most abundant form of biomass and a potential replacement of crude oil. Unlike today's oil refineries, future bio-refineries will fractionate biomass rather than crude oil. This paper presents developments in catalysis and process engineering for the conversion of lignocellulose into energy carriers, alternative fuels and chemicals for future bio-refineries.

Keywords Bio-refinery · Biomass gasification · Biomass to chemicals · Biofuels

1 Introduction

The chemical and physical structure of lignocellulosic biomass has naturally evolved to resist chemical and biological attack, making it extremely hard to break down. Over the last decade we have developed catalytic processes which combine thermo-chemical, catalytic, and enzymatic methods to breakdown and convert lignocellulose into platform chemicals like sorbitol, synthesis gas, 5-hydroxymethylfurfural and levulinic acid. Several novel concepts have been developed over the last few years which can lower the cost of processing ligno-cellulose biomass and minimise separation costs. This paper will present our latest results from the following works.

Reactive flash volatilisation (RFV) process was developed to directly convert lignocellulose biomass into char and tar free synthesis gas in a millisecond residence time catalytic reactor, which produces controlled CO to H_2 ratio [1]. Pinewood and eucalyptus sawdust were used as feedstock and operating parameters such as carbon to oxygen feed ratio, and carbon to steam feed ratio were tested for

A. Tanksale (✉)
Department of Chemical Engineering, Catalysis for Green Chemicals Group,
Monash University, Clayton, VIC 3800, Australia
e-mail: akshat.tanksale@monash.edu

K.V. Raghavan and P. Ghosh (eds.), *Energy Engineering*,
DOI 10.1007/978-981-10-3102-1_17

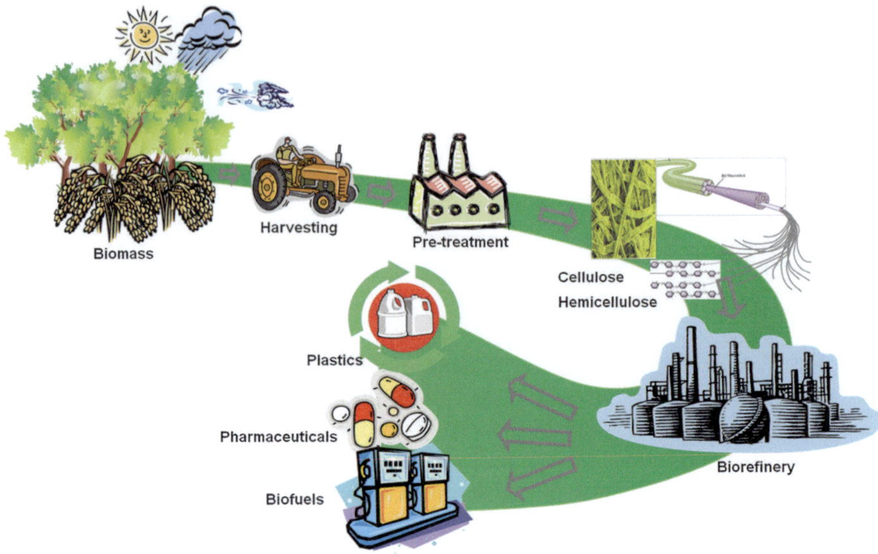

Fig. 1 Integrated Bio-refinery concept which converts wastes from agriculture and forestry industries into liquid fuels, gaseous fuels (such as synthesis gas) and chemicals

various nickel based catalysts. High gasification efficiency was observed in the pinewood RFV with Re-Ni and Rh-Ni catalysts.

The synthesis gas as produced was converted into formaldehyde using a novel method in which equimolar quantities of carbon monoxide and hydrogen in a catalytic slurry phase reaction [2]. Thermodynamics analysis showed that the reaction is equilibrium limited in the gas phase whereas it is kinetically limited in the liquid phase.

Novel enzymatic pretreatment has been developed which integrates mild liquid hot water pretreatment followed by enzymatic digestion using xylanases, novel ferulic acid esterases on corn stover, wheat straw and eucalyptus. The pretreated biomass was converted to value added chemicals such as 5-hydroxymethylfurfural and furfural by using levulinic acid as a catalyst and compared against untreated biomass.

Integration of these processes in a chemical industry will result in simultaneous production of energy, fuels and value added chemicals in a true bio-refinery concept (Fig. 1).

References

1. F.L. Chan, A. Tanksale, Catalytic steam gasification of cellulose using reactive flash volatilization. Chem. Cat. Chem. **6**, 2727 (2014)
2. A.M. Bahmanpour, A. Hoadley, A. Tanksale, Formaldehyde production via hydrogenation of carbon monoxide in the aqueous phase. Green Chem. **17**, 3500 (2015)

Sulfonated Polyethersulfone/Torlon Blend Membrane Incorporated with Multiwalled Carbon Nanotubes for Energy Production from Kitchen Wastewater Using Microbial Fuel Cell

Harsha Nagar, G. Anusha and S. Sridhar

Abstract Microbial fuel cell (MFC) is extensively considered as the most efficient alternative energy source which uses organic wastes to generate both electricity and clean water. In this study, an indigenously synthesized sulfonated polyethersulfone/Torlon blend membrane incorporated with multiwalled carbon nanotubes was characterized for ion exchange capacity (IEC 1.45 meq/g) and proton conductivity (0.86 Scm^{-1}) at 30 ± 5 °C. A high power density of 190.73 mW m^{-2} was achieved at a corresponding current density of 260.20 mA m^{-2} along with significant COD removal efficiency of 87.5 % from kitchen wastewater.

Keywords Microbial fuel cell · Kitchen wastewater · Mixed culture · Proton conducting membrane, Power density

1 Introduction

Effluent treatment for water reclamation could be economical and attractive if power could be simultaneously generated from the waste stream. Microbial fuel cells (MFC) are clean energy devices which exploit the metabolism of a microorganism to generate bio-energy by catalyzing the oxidation of fuels along with treatment of the wastewater [1]. In this study, MFC was operated with equal volumes of 0.6 L in both anode and cathode compartments that were separated by a selectively permeable novel proton exchange membrane (PEM). This indigenous proton conducting blend membrane was employed for treatment of kitchen effluent as shown in Fig. 1. In the anodic compartment, a sludge of kitchen wastewater was pretreated in anaerobic mode wherein mixed bacterial culture (cow dung) was added. The presence of bacteria as biocatalyst in the anodic cell oxidized the

H. Nagar · G. Anusha · S. Sridhar (✉)
Membrane Separations Group, Chemical Engineering Division,
CSIR-Indian Institute of Chemical Technology, Hyderabad 500007, India
e-mail: sridhar11in@yahoo.com

K.V. Raghavan and P. Ghosh (eds.), *Energy Engineering*,
DOI 10.1007/978-981-10-3102-1_18

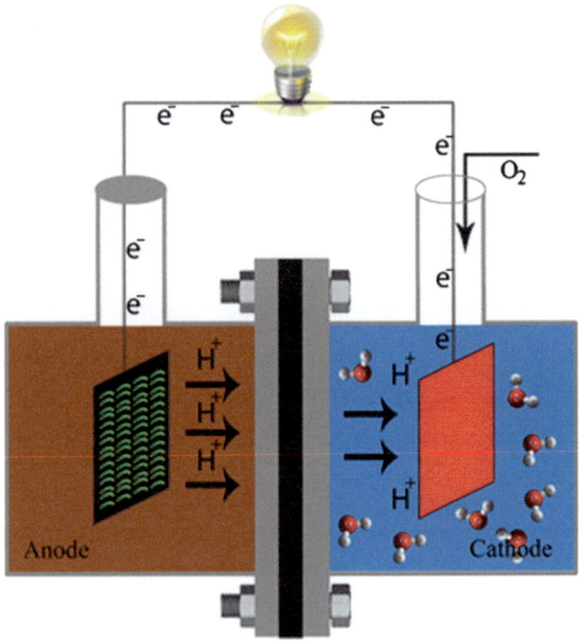

Fig. 1 MFC experimental setup

substrates to generate electrons and protons. These electrons get transported through an external circuit from the anode side to the cathode side. Protons pass through the membrane and enter the aerated cathode cell to combine with oxygen and produce clean water as byproduct. PEMs are one of the important factors that influence the performance of MFCs due to their role in transportation of proton across the electrodes. The membrane also reduces the diffusivity of oxygen and transportation of other substrates that affect the MFC performance. Oxygen crossover into the anode chamber causes an aerobic condition resulting in a decline in power density and effluent treatment efficiency, along with diffusion of media to the cathode side [2]. In this study, a dense sulfonated polyethersulfone/Torlon blend membrane incorporated with multiwalled carbon nanotubes was synthesized using solution casting and total solvent evaporation technique. The performance of the blend membrane containing inorganic filler was evaluated in terms of efficiency of wastewater treatment and power generation capacity.

2 Experimental Results

Kitchen wastewater collected from CSIR-IICT canteen contained moderately high COD value of 2400 ppm that was used as substrate for the mixed bacterial culture. After treatment of the wastewater using MFC system and completion of one cycle of operation, the COD could be reduced to 300 ppm. The results revealed the COD

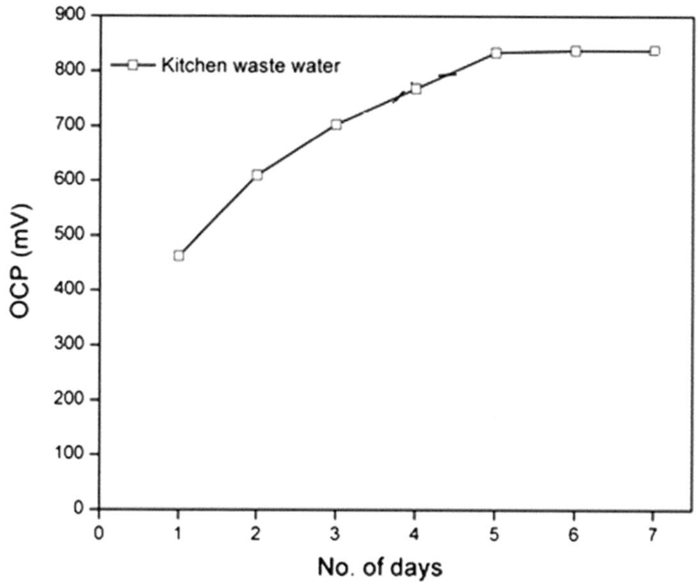

Fig. 2 OCV graph of kitchen wastewater

removal efficiency to be 87.5 %, and the treated water could be further processed by nanofiltration or reverse osmosis to produce water of recyclable quality. This shows that the microorganisms were able to degrade the organic matter present in the feed to a large extent at a high rate.

Figure 2 represents the open circuit voltage (OCV) graph of kitchen wastewater substrate in a condition wherein no current is flowing. At the beginning, the voltage was observed to gradually increase and after a few days of operation it became constant. The MFC operation with kitchen wastewater reached the maximum OCV value of 826 ± 3 mV after one week of inoculation. Upon attaining the maximum voltage, the measurement of polarization curve of different substrates under an external resistance in the range 1 Ω–12 kΩ was performed. Figure 3 shows the cell voltage graph of the kitchen effluent, with the drop in cell voltage being influenced by losses in activation energy besides changes in ohmic resistance and mass transfer rates. The current generation shows an inverse relation with external resistance. A peak power output of 190.73 mW m^{-2} could be observed at a current density of 260.20 mA m^{-2} and external resistance of 100 Ω from the polarization curve. The high power density and low voltage drop at reduced external resistance indicate lower mass transfer rates at the electrodes. From the data, it is observed that waste degradation efficiency of the microbes is high at low resistance when compared to case involving greater resistance wherein microbes have to perform the function of transferring electrons to the anode once they are released in the closed circuit. The reason behind the sudden drop in the voltage at lower resistance is due to rapid exhaustion of the substrate because of which electrons are unable to reach the

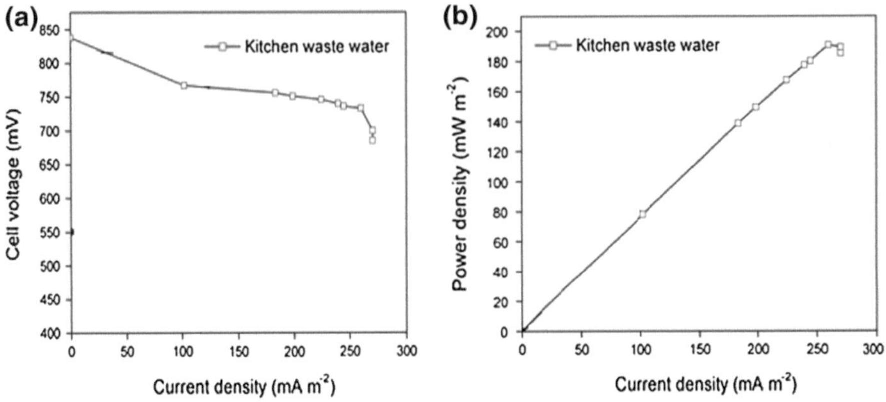

Fig. 3 **a** Cell voltage. **b** Power density curve of membrane

cathode. Swift voltage drop was observed in all MFCs reported in literature at high external resistance which is a result of the low energy available for initiating the redox reaction that enables transportation of electrons from the microbe cell terminal to the anode surface [3]. The high power density and COD removal efficiency of the present blend membrane is found comparable to that of commercial Nafion117 membrane which exhibited corresponding values of 30.99 mW m^{-2} with 75 % COD removal for a similar anaerobic effluent from another effluent treatment plant (ETP) [4].

3 Conclusions

Treatment of kitchen wastewater with added benefit of biologically generated electricity appeared feasible using microbial fuel cell. The sulfonated polyethersulfone/Torlon blend membrane incorporated with multiwalled carbon nanotubes (SPES/Torlon-MWCNT) was effective for proton conduction after using mixed culture (cow dung) for breaking down the COD molecules present in the effluent. The mixed culture exhibited potential for MFC research due to their higher metabolic activity which facilitated enhanced power density of 190.73 mW m^{-2} along with substantial water reclamation capability owing to the high efficiency of 87.5 % in COD removal. The results suggest the feasibility to produce alternate membranes that are cost effective but capable of exhibiting performance comparable to state of the art Nafion membranes.

References

1. P. Aelterman, K. Rabaey, H.T. Pham, N. Boon, W. Verstraete, Continuous electricity generation at high voltages and currents using stacked microbial fuel cells. Environ. Sci. Technol. **40**, 3388 (2006)
2. M. Ghasemi, E. Halakoo, M. Sedighi, J. Alamd, M. Sadeqzadeh, Performance comparison of three common proton exchange membranes for sustainable bioenergy production in microbial fuel cell. Procedia CIRP **26**, 162 (2015)
3. A. Nandy, V. Kumar, S. Mondal, K. Dutta, P.P. Kundu, Performance evaluation of microbial fuel cells: effect of varying electrode configuration and presence of a membrane electrode assembly. New Biotechnol. **32**, 272 (2015)
4. S. Fatemi, A.A. Ghoreyshi, G. Najafpour, M. Rahimnejad, Bioelectricity generation in mediator-less microbial fuel cell application of pure and mixed cultures. Iranica J. Energy Environ. (IJEE) **3**, 104 (2012)

Development of Sulfonated Polyethersulfone/Matrimid Acid-Base Blend Membrane for Energy Production Through Fuel Cells

Harsha Nagar, G. Anusha and S. Sridhar

Abstract Expansion in population and exhaustion of non-renewable energy sources requires exploration of other power resources such as fuel cells that have emerged as an alternative clean and green energy. The work presents the performance of an indigenously prepared blend made of sulfonated polyethersulfone (SPES) and Matrimid polyimide for membrane based FCs. Blending induced higher tensile strength and IEC (1.3 meq g^{-1}) with proton conductivity in the range 0.068–0.12 S cm^{-1} over a temperature range of 25–80 °C respectively. Methanol permeability of the membrane was 7.46×10^{-8} cm^2 s^{-1} which is lower than Nafion 117 (25.0×10^{-7} cm^2 s^{-1}). The power density of the membrane was 0.0172 W cm^{-2} at 25 ± 5 °C, thus exhibiting potential for PEMFC and DMFC applications.

Keywords Fuel cell · Acid-base blend · Methanol permeability · Proton conductivity

1 Introduction

Increasing population causes high energy consumption and global warming has resulted in depletion of non-renewable energy resources at a rapid rate. Therefore there is a global need to identify alternative energy source which is environmentally benign. Membrane based fuel cells have the potential to become an important energy conversion technology. The essential component of the fuel cell (FCs) is the polymer electrolyte membrane (PEM), which act as a barrier to avoid direct contact between the fuel and oxidant and work as a proton conducting medium [1]. Generally, Nafion, a perfluorinated membrane, manufactured by Dupont is used as the commercial membrane in fuel cells. However, expensive catalyst, low thermal

H. Nagar · G. Anusha · S. Sridhar (✉)
Membrane Separations Group, Chemical Engineering Division, Indian Institute of Chemical Technology, Hyderabad 500007, India
e-mail: sridhar11in@yahoo.com

K.V. Raghavan and P. Ghosh (eds.), *Energy Engineering*,
DOI 10.1007/978-981-10-3102-1_19

and mechanical strength of membrane, poor water retention and high methanol crossover at higher temperatures, limits Nafion's usage and hinder the widespread commercialization of fuel cells [2]. Consequently, significant research has been done to amend the Nafion membrane or develop new polymer electrolytes with desired properties for fuel cell applications. Over the years membranes have been modified either by sulfonation or blending with other polymers. In the present work, an acid base blend of sulfonated polyethersulfone (SPES) and matrimid is reported. The performance of the membrane was evaluated under different parametric conditions.

2 Experimental Results

Ion exchange capacity (IEC) is useful in quantifying the number of unreacted functional groups available in the membrane for proton conduction as provided in Table 1. The IEC value of SPES/Matrimid was found to be 1.3 meq g^{-1}. A high IEC value ensures easy proton conduction through the membrane. Methanol crossover slows down the cathode side reaction and leads to fuel losses in DMFC. Table 1 represents methanol permeability values of all the membranes at ambient temperature (25 ± 5 °C). Methanol transportation occurs through ionic channels and hydrophilic sites in the membrane. The interaction between sulfonated and imide groups of the blend narrows down the path of methanol permeation which restricts methanol crossover.

Proton conductivity is an important characteristic of fuel cell membrane and mainly depends upon the temperature and hydrophilic nature of membrane. Table 1 exhibits proton conductivity of the membrane at hydrated condition and ambient temperature (30 ± 5 °C). The proton conductivity of SPES/Matrimid is directly proportional with temperature as shown in Fig. 1. An increase in temperature provides more energy for proton diffusion due to favorable thermodynamic conditions of lower activation energy requirement as per Arrhenius law, wherein a straight line is the resultant plot with the gradient representing the activation energy as shown in Fig. 2. The activation energy of SPES/Matrimid membrane is 9.28 kJ/mole comparable to commercial Nafion 117 (9.56 kJ/mol) [3]. The highest conductivity of the SPES/Matrimid membrane is observed to be 0.12 S cm^{-1} at a temperature of 80 °C. Selectivity is the ratio of proton conductivity and methanol permeability useful for

Table 1 Membrane properties at ambient temperature (25 ± 5 °C)

Membranes	Proton conductivity	Methanol permeability	Selectivity	IEC
PES	0.0023	1.23×10^{-7}	2.11×10^{5}	0.61
SPES	0.031	1.26×10^{-7}	1.59×10^{5}	1.23
Matrimid	0.049	3.26×10^{-7}	1.50×10^{5}	0.70
PES/Matrimid	0.033	1.13×10^{-7}	2.90×10^{4}	0.67
SPES/Matrimid	0.068	7.46×10^{-8}	9.11×10^{5}	1.3

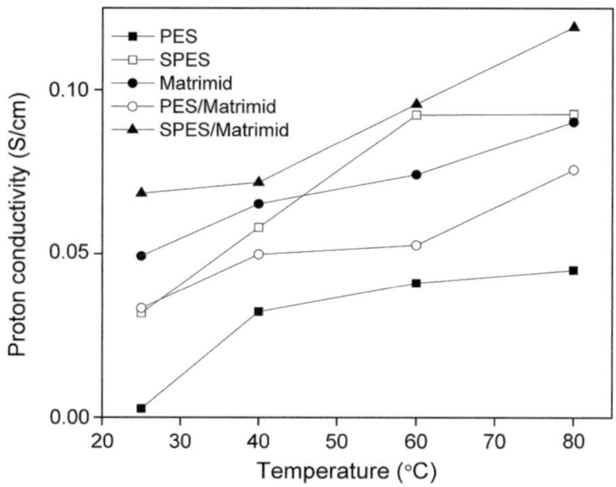

Fig. 1 Proton conductivity versus temperature

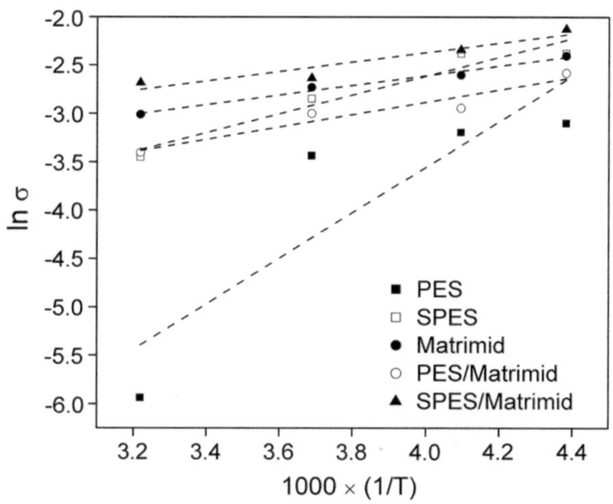

Fig. 2 Arrhenius graph

DMFC performance. of the SPES/Matrimid membrane is 9.11×10^5 Ss cm^{-3} which is comparable to Nafion 117 membrane (4.05×10^4 Ss cm^{-3}) [4]. The performance of single unit fuel cell equipped with PES/Matrimid and SPES/Matrimid membranes at ambient temperature (25 ± 5 °C) was studied. Due to the presence of activation, ohmic and mass transfer resistances, the cell voltage of the membranes shows an inverse relationship with the resistances as shown in Fig. 3a. The high ionic conductivity of SPES/Matrimid enables a high power density of 17.2 mW cm^{-2} as depicted in Fig. 3b which was higher compared to other membranes. The results clearly emphasizes that the performance of the indigenous SPES/Matrimid blend membrane is ideal for fuel cell application.

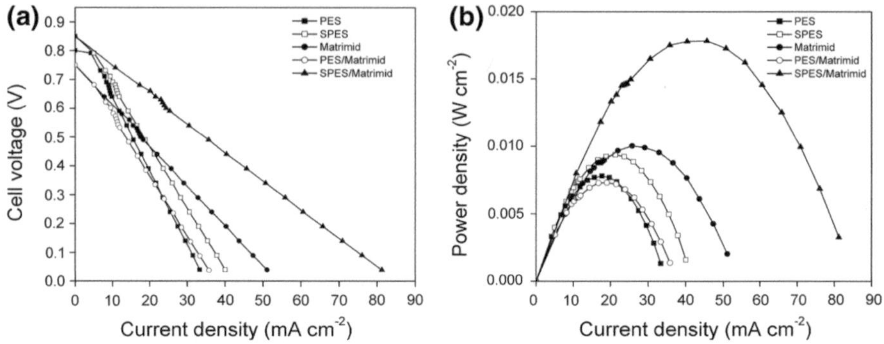

Fig. 3 **a** Cell voltage. **b** Power density of membranes

3 Conclusions

In a nutshell it can be concluded that the blend membrane synthesized from sulfonated polyethersulfone and matrimid has optimum water uptake and low methanol permeability due to ionic interactions between imide and sulfonic acid groups and intermolecular hydrogen bonding. The blending ratio of SPES/Matrimid (50:50) is more suitable due to its high stability compare to other ratios. The high proton conductivity comparable to commercial Nafion 117 along with good selectivity makes this acid-base blend membrane useful for fuel cell applications at high temperature.

References

1. Z. Zicheng, F. Yongzhu, M. Arumugam, Novel blend membranes based on acid-base interactions for fuel cells. Polymers **4**, 1627 (2012)
2. N. Vladimir, M. Jonathan, W. Haijiang, Z. Jiujun, A review of polymer electrolyte membranes for direct methanol fuel cells. J. Power Sources **169**, 221 (2007)
3. A. Kausar, M. Khurram, M. Siddiq, Sulfonated poly(sulfone-pyridine-amide)/sulfonated polystyrene/multiwalled carbon nanotube-based fuel cell membranes. Poly. Eng. Sci **55**, 1776 (2015)
4. H.S. Thiam, W.R.W. Daud, S.K. Kamarudina, A.B. Mohamad, A.A.H. Kadhuma, K.S. Loh, E. H. Majlana, Nafion/Pd–SiO_2 nanofiber composite membranes for direct methanol fuel cell application. Int. J. Hydro. Ener **38**, 9474 (2013)

Adaption of a Green Technology (Friction Stir Welding) to Join Fusion Energy Materials

Vijaya L. Manugula, Koteswararao V. Rajulapati, G. Madhusudhan Reddy and K. Bhanu Sankara Rao

Abstract Traditional welding techniques result in the creation of large heat affected zones (HAZ) as well as several filler materials need to be used. In addition, evolution of several unwanted gases into the atmosphere as well as solidification of some unwanted phases in the materials being welded demands for the alternate welding routes by which all these problems could be avoided. As an attempt in this direction, friction stir welding has been carried out on 6 mm thick reduced activation ferritic martensitic steel at various rotational speeds. The resultant microstructures and the observed properties are discussed in this paper.

Keywords FSW · INRAFM steel · Heat affected zones · Welding · HAZ

1 Introduction

Friction stir welding (FSW), a solid state welding process which is invented at the welding institute (TWI) Cambridge U.K. in 1991 has shown great potential for joining of low strength materials like aluminium, copper and magnesium alloys [1]. Being a solid state process FSW eliminates many of the solidification related defects associated with fusion welding processes. FSW is an environmentally

V.L. Manugula · K.V. Rajulapati (✉)
School of Engineering Sciences and Technology, University of Hyderabad,
Hyderabad 500046, India
e-mail: kvrse.voh@gmail.com

V.L. Manugula
Department of Metallurgical and Materials Engineering, Mahatma Gandhi Institute of Technology, Hyderabad 500075, India

G. Madhusudhan Reddy
Defence Metallurgical Research Laboratory, Hyderabad 500050, India

K. Bhanu Sankara Rao
Ministry of Steel (Govt. of India), Mahatma Gandhi Institute of Technology, Hyderabad 500075, India

K.V. Raghavan and P. Ghosh (eds.), *Energy Engineering*,
DOI 10.1007/978-981-10-3102-1_20

friendly process due to no arc radiation so no toxic fumes, smoke and hazardous waste produced during welding when compared to other conventional fusion welding processes. FSW process is energy efficient process and higher travel speeds could be achieved than those attained with arc welding processes [2]. In FSW process a non-consumable rotating tool is inserted into the work piece which creates heat due to friction and plastic deformation that leads to form plasticised zone around the tool. The material is transported from advancing side to retreating side.

Reduced activation ferritic martensitic steel (RAFM) occupies an unique status as structural engineering material for fusion reactors by virtue of their swelling resistance and excellent thermal properties [3]. Since RAFM steels are structural materials for manufacture of fusion reactors; weldability has been an important consideration. The fabrication of the TBM structure necessitates development of advanced joining methods as high heat input in the conventional fusion welding process like SMAW and TIG produces wider heat affected zone (HAZ) and intercritical zones which lead to premature failure in the HAZ of welded joints at high temperatures due to the occurrence of Type IV cracking The formation of δ-ferrite in the fusion zone during solidification could not be avoided in conventional and advanced fusion welding process which deteriorates the mechanical properties. FSW is adapted for the welding of RAFM steel as these welds are free from solidification related problems and also minimise the HAZ. Therefore the present investigation is aimed at establishing the feasibility of joining 6 mm thick Indian reduced activation ferritic martensitic steel (INRAFM) by FSW process and explore the phase transformation that occurs during FSW.

2 Experimental Result

2.1 Experimental Procedure

The material used in this study is INRAFM steel (Fe (bal.)-9Cr-1.4W-0.06Ta-0.23V in wt%) was produced by Midhani, Hyderabad, India. Bead-on-plate FSW was conducted on 6 mm thick plate (Normalised and Tempered) using PcBN tool with Argon gas shielding to prevent the oxidation of the surface during welding. The interface temperature was measured by thermography, CEDIP Infrared Camera was operated with Indium-Antimony detector, frame rate of 50–380 Hz and accuracy less than 25 mK at 298 K. FSW was carried out at low and high rotational speeds (200 and 700 rpm) and at constant traverse speed of 30 mm/min.

The samples for FEG-SEM were extracted from FSW plates, prepared according to standard metallographic procedures, and etched with Villela's reagent. Thin foils of 100 mm were obtained from the central regions of the SZ and 3 mm discs were punched. These discs were electro polished in a solution containing 90 % methanol and 10 % perchloric acid, at 15 V and 250 K, in a twin jet apparatus and were examined in an electron microscope operating at an acceleration voltage of 200 keV.

Vickers micro hardness test was performed at the mid thickness of the cross section across the SZ with 300 g load at regular interval of 0.5 mm distance and dwell time 20 s. Charpy sub size samples 55 × 10 × 5 mm were extracted by EDM wire cut according to ASTM E-23 standards with notch machined at SZ centre.

2.2 *Results and Discussion*

Thermography results clearly revealed that the SZ temperature for 200 rpm was below A_{c1} temperature for this steel and it was above A_{c3} temperature for 700 rpm [4].

The macrostructure of the FSW joint consists of weld nugget, thermo mechanically affected zone (TMAZ), and BM. Higher heat input lead to widening of SZ. BM microstructure in the normalised and tempered condition is shown in Fig. 1a consists of tempered martensite with lath and prior austenite grain boundaries (PAGBs) decorated by $M_{23}C_6$ type carbides rich in Cr while the fine interlath precipitates corresponded to Ta, V rich MX type precipitates. The SZ

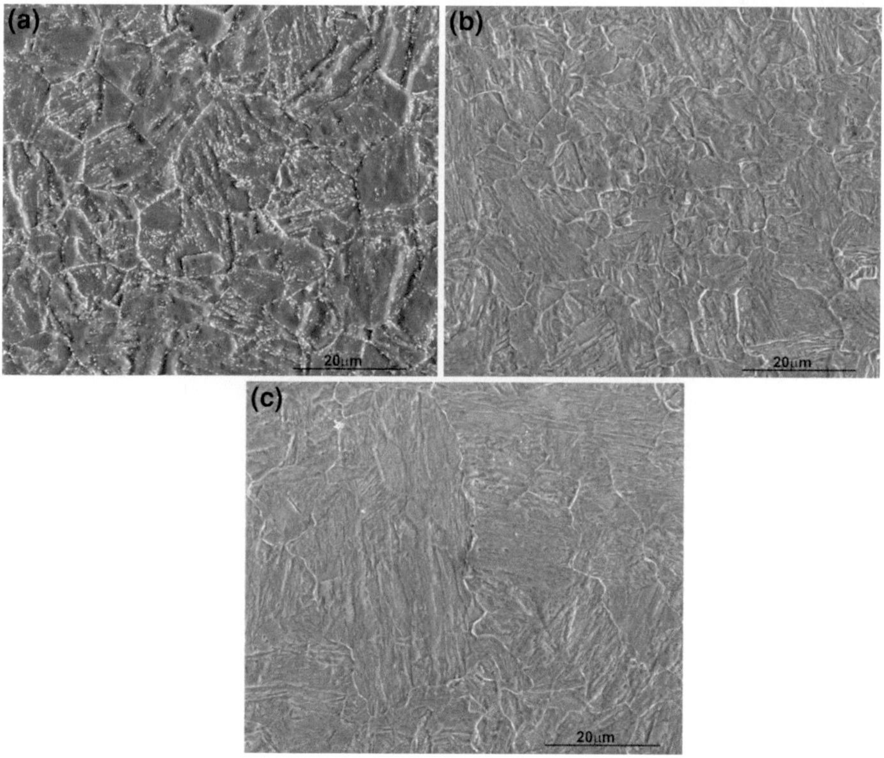

Fig. 1 SEM microstructures of **a** Base metal **b** 200 rpm Stir zone centre and **c** 700 rpm stir zone centre

microstructure of the 200 rpm as shown in Fig. 1b reveals fine scale distribution of carbides and grain refinement due to the occurrence of dynamic recrystallization as the temperature during FSW is below transformation temperature (Ac1 of this steel is 817 °C). The interface temperature in 700 rpm condition measured by thermography was above A_{C3}. High temperatures caused the dissolution of carbides and led to grain growth during FSW as shown in Fig. 1c.

From TEM investigation it was observed that very fine needle type precipitates were formed in SZs of low and high rotational speeds and these correspond to the stoichiometry of Fe_3C. The 200 rpm condition revealed the presence of primary un dissolved TaC precipitates and very fine carbides in the matrix of SZ. The SZ microstructures of 700 rpm were composed of freshly formed martensite plates.

Vickers micro hardness profile of the FSW joints is shown in Fig. 2. Hardness within the SZs is higher than the BM. The high hardness in the SZ for 200 rpm condition may be related to fine grain size and very fine un dissolved carbides. The high hardness in the SZ for 700 rpm condition could be related to un tempered martensite.

Charpy impact toughness was evaluated for the BM and FSW joints and the results were converted to standard size using a scaling factor of 3/2. The Charpy impact toughness for BM is 156 J and the FSW joint for 200 rpm condition is 150 J close to BM and for 700 rpm condition the welds exhibited poor toughness i.e., as low as 13.5 J. The reason for low toughness may be attributed to coarse grain, and un tempered martensite in the SZ. The high toughness for 200 rpm condition is related to refinement of grain size, fine scale distribution of undissolved carbides and absence of phase transformation. All these findings are comprehensively discussed in Ref. [5].

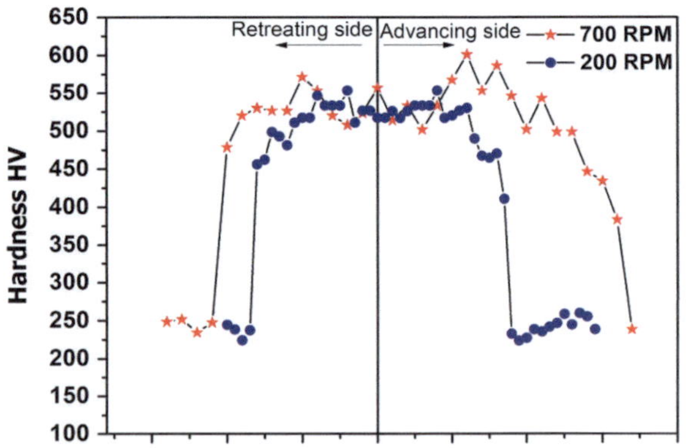

Fig. 2 Vickers microhardness profile at mid thickness across the stir zone

3 Conclusions

FSW below Transformation temperature was successfully produced on 6 mm thick RAFM steel plate. Grain boundary carbides dissolved and Fe_3C precipitated in low and high rotational speeds. The impact toughness of the 200 rpm is close to that of BM due to non occurrence of martensite as the peak temperatures seen in SZ was below Ac_1 temperature. Impact toughness drastically reduced for 700 rpm weld due to the occurrence of martensite. PWHTs are necessary to bring down the hardness in SZ and restoration of grain and lath boundary carbides.

References

1. W.M Thomas, E.D. Nicholas, J.C. Needham, M.G. Murch, P. Temple-Smith, G.J. Dawes, Friction stir butt welding (1991), GB Patent No. 9125978.8, International Patent No. PCT/GB92/02203
2. Innovations in materials manufacturing fabrication and environmental safety", ed. by Mel Schartz (CRC press, Taylor and Francis)
3. R.L. Klueh, D.S. Gelles, T.A. Lechtenberg, Development of ferritic steels for reduced activation; The U.S. program. J. Nucl. Mater. **141–143**, 1081–1087 (1986)
4. A. Ravi Kiran, R. Mythili, S. Raju, S. Saroja, T. Jayakumar, E. Rajendra Kumar, Decomposition modes of austenite in 9Cr–W–V–Ta reduced activation ferritic–martensitic steels. Mat. Sci. Tech. **31**, 448–459, (2015)
5. V.L. Manugula, K.V. Rajulapati, G.M. Reddy, R. Mythili, K.B.S. Rao, A critical assessment of the microstructure and mechanical properties of friction stir welded reduced activation ferritic-martensitic steel. Mater. Des. **92**, 200–212 (2016)

Advanced Materials for Indian Future Nuclear Energy Systems

G.V. Prasad Reddy, R. Sandhya and K. Laha

Abstract Development of materials with enhanced high temperature mechanical properties is vital for Indian future nuclear energy programme. Indian efforts towards the development of nitrogen enhanced 316LN SS structural steel and 9Cr-ODS fuel cladding steel for fast reactor, and reduced activation ferritic-martensitic (RAFM) steel for blanket module of fusion reactor are described.

Keywords SFR · Fusion reactor · 316LN SS · 9Cr-ODS · RAFM steels

1 Introduction

In India, sodium cooled fast reactor (SFR) is envisaged as a clean source of energy along with fusion reactor. For economic competitiveness of SFR, Indira Gandhi Centre for Atomic Research has undertaken a programme to develop high temperature materials. Materials being developed for SFR are high nitrogen 316LN SS for reactor internals to extend reactor life, and 9Cr-ODS steel fuel clad to increase burnup. For fusion reactor program, India specific RAFM steel has been developed for blanket module facing intense neutron irradiation at high temperature.

2 Mechanical Behavior of Structural Materials

The 316LN SS with different nitrogen contents have been studied for optimum combination of mechanical properties. Tensile and creep strengths of the steel increase with nitrogen content; whereas fatigue and creep-fatigue interaction (Fig. 1) studies indicate optimum nitrogen content around 0.12 wt%. Nitrogen enhanced 316LN SS with nitrogen content in the range 0.11–0.13 wt% has been

G.V. Prasad Reddy (✉) · R. Sandhya · K. Laha
Metallurgy Programme, Indira Gandhi Centre for Atomic Research, Kalpakkam, India
e-mail: prasadreddy@igcar.gov.in

K.V. Raghavan and P. Ghosh (eds.), *Energy Engineering*,
DOI 10.1007/978-981-10-3102-1_21

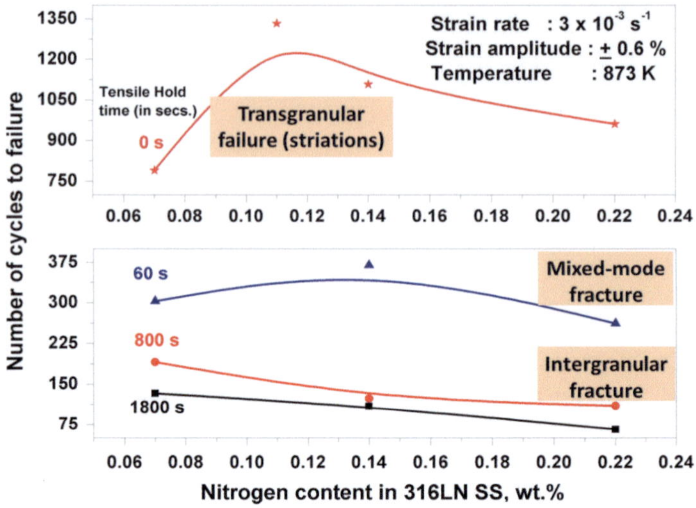

Fig. 1 Effect of nitrogen on creep-fatigue interaction behaviour of 316LN steel

Fig. 2 Creep rupture strength of developed 9Cr-ODS steel, compared with other steels

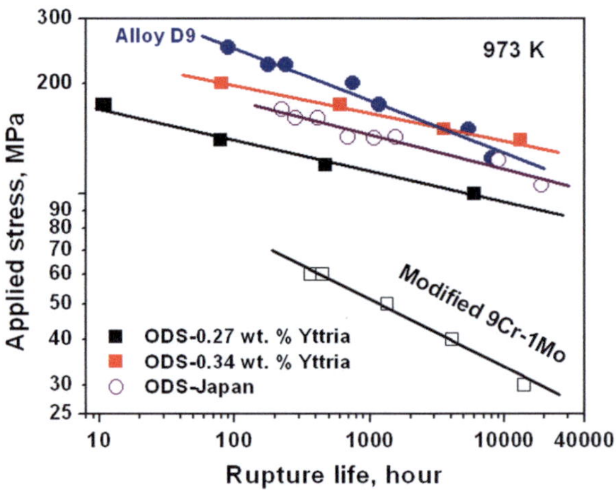

optimized to enhance SFR life. Inherently void swelling resistant ferritic steel has been considered as fuel clad to increase burnup. The challenge for adequate creep strength of ferritic steel has been met upon oxide dispersion strengthening (ODS) the steel by Yttria nano particles. The technology to produce thin walled 9Cr-ODS cladding tube has been realized through powder metallurgy route followed by extrusion and cold pilgering. Creep strength of the 9Cr-ODS tube is comparable to the presently used alloy D9 (Fig. 2). With the participation in ITER program, India is vigorously pursuing harnessing energy through nuclear fusion. One of the challenges is the development of plasma facing RAFM steel. The

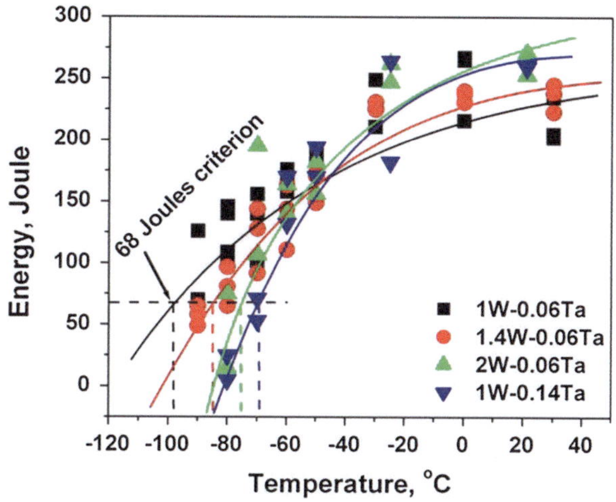

Fig. 3 Effect of W and Ta on DBTT of 9Cr-RAFM steel

material is needed to have very low ductile to brittle transition temperature (DBTT) with adequate tensile, creep and fatigue strength. In the internationally developed 9Cr-RAFM steels, the tungsten and tantalum contents have been optimised for better combination of DBTT, tensile, fatigue, creep and creep-fatigue interaction to develop India specific RAFM steel (INRAFM) having DBTT less than −70 °C (Fig. 3).

3 Conclusions

The indigenously developed INRAFM steel, Nitrogen Enhanced 316LN SS, and highly radiation-resistant 9Cr-ODS steel shows immense potential for use in Indian nuclear energy systems.

Reference

1. V. Shankar, K. Mariappan, A. Nagesha, G.V. Prasad Reddy, R. Sandhya, M.D. Mathew T. Jayakumar, Effect of tungsten and tantalum on the low cycle fatigue behavior of reduced activation ferritic/martensitic steels. Fusion Eng. Des. **87**, 318 (2012)

Mitigating Embrittlement in Structural Steels of Power Plants by Grain Boundary Engineering

T. Karthikeyan, Saroja Saibaba and M. Vijayalakshmi

Abstract The Cr-Mo alloyed steels used as structural materials of power plants, are susceptible to grain boundary embrittlement upon long-term exposure at service temperatures. The present study is aimed at characterization of grain boundary nature in the 9Cr-1Mo-0.1C steel, and identification of process treatment for mitigating the temper embrittlement problem. A new heat treatment of DNT (Normalizing by Double austenitization and Tempering) was designed to refine the prior-austenite to grain size from 25 to 12 μm. As compared to CNT (Conventional Normalizing & Tempering) treatment, the DNT treated steel exhibited lower DBTT (Ductile-to-Brittle Transition Temperature), comparable room temperature and high temperature mechanical properties, while the embrittlement effect after ageing at 773 K for 5,000 h was less pronounced.

Keywords High-Cr steels · Grain refinement · Temper embrittlement

1 Introduction

The ferritic/martensitic steels (such as the 9Cr-1Mo-0.1C steel) are used as structural materials in power plants, and are considered for nuclear reactor core applications due to their excellent radiation resistance. The embrittlement phenomena that lead to reduction in toughness property upon long-term service at high temperatures and irradiation environments, is an important design issue to be tackled for ensuring performance of structural components. Loss of grain boundary cohesive strength (due to segregation of impurity elements or precipitation of brittle phases along high interfacial energy boundaries) is an important recognized cause for embrittlement of steels, and *Grain Boundary Engineering* presents a viable approach for mitigating the embrittlement problem.

T. Karthikeyan (✉) · S. Saibaba · M. Vijayalakshmi
Physical Metallurgy Group, Indira Gandhi Centre for Atomic Research, Kalpakkam, India
e-mail: tkarthi@igcar.gov.in

K.V. Raghavan and P. Ghosh (eds.), *Energy Engineering*,
DOI 10.1007/978-981-10-3102-1_22

2 Experiments

The steel blocks were subjected to two types of heat treatments, namely **CNT**—Conventional Normalizing (1323 K/1 h/Air cool) and Tempering (1023 K/2 h/air cool), and **DNT**—Normalizing by Double austenitization (1323 K/1 h/water quench, 1233 K/1 h/air cool) and Tempering (1023 K/2 h/air cool). The microstructures and grain boundary character distribution of the samples were characterized by optical microscopy and SEM-EBSD techniques. The room temperature and high temperature (823 K) mechanical properties of Yield strength/Tensile strength/Ductility were evaluated, and the ductile-to-brittle transition behavior was studied from Charpy impact toughness measurements. Also, the CNT & DNT steels were exposed to long-term thermal aging (823 K/5000 h), and the Charpy tests were repeated to evaluate the temper embrittlement behavior.

3 Results

A homogeneous tempered martensite was obtained after CNT and DNT treatments, but the average prior-austenite grain size was nearly halved (12 μm) for DNT treated steel. The grain boundary character distributions were similar with the amount of 'coincidence site lattice boundaries' (Σ1–29) of 35 % [1]. Both CNT and DNT treated steels exhibited similar hardness (250 HV10), room temperature (540 MPa/700 MPa/30 %), and 823 K (340 MPa/410 MPa/>21 %) tensile properties, while the 823 K ductility was higher (29 %) for DNT treated steel. The ductile-to-transition curves of the two heat treated steel is shown in Fig. 1.

The DNT treated steel exhibited a lower DBTT (by 15 K) and higher Upper Shelf Energy (by 35 Joules), while a similar shift in DBTT upon thermal aging occurred in both heat treated steels. Fractography revealed fibrous pull-outs for high

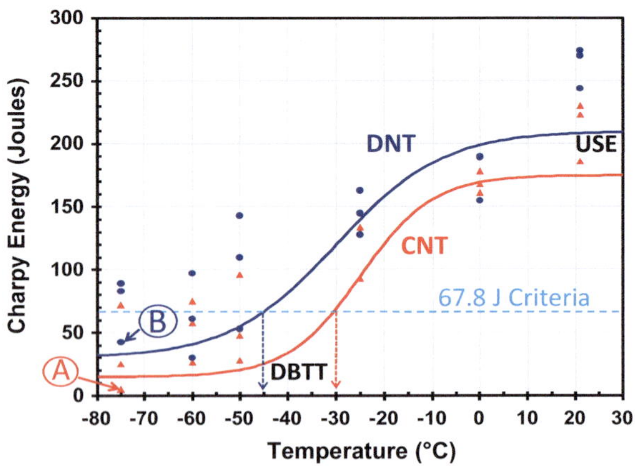

Fig. 1 Comparison of Charpy impact toughness as a function of test temperatures in 9Cr-1Mo steel subjected to CNT and DNT treatments

temperature tests, and transgranular cleavage facets for specimens broken at brittle temperature regimes. Brittle fractured (at −75 °C) samples (**A** of DNT with 42.5 Joules versus **B** of CNT with 5 Joules) exhibited higher crack tortuosity with fractal dimension increment higher in **A** (0.180) compared to **B** (0.128).

4 Conclusion

The DNT heat treatment of 9Cr-1Mo steel resulted in homogeneous tempered martensite microstructure with smaller grain size and 35 % CSL boundaries, showed adequate tensile properties, better impact toughness properties and resistance to temper embrittlement, and with brittle fractured specimen exhibiting higher fractal dimension, as compared to CNT treated steel.

Reference

1. T. Karthikeyan, M.K. Dash, S. Saroja, M. Vijayalakshmi, Metall. Mater. Trans. A **44**, 1673–1685 (2013)

Author Index

K.V. Raghavan and P. Ghosh (eds.), *Energy Engineering*,
DOI 10.1007/978-981-10-3102-1

MIX
Papier aus verantwortungsvollen Quellen
Paper from responsible sources
FSC® C105338

If you have any concerns about our products,
you can contact us on
ProductSafety@springernature.com

In case Publisher is established outside the EU,
the EU authorized representative is:
Springer Nature Customer Service Center GmbH
Europaplatz 3, 69115 Heidelberg, Germany

Printed by Libri Plureos GmbH
in Hamburg, Germany